Researching functional ecology
in Kosciuszko National Park

Researching functional ecology in Kosciuszko National Park

Edited by Hannah Zurcher, Ming-Dao Chia, Michael Whitehead and Adrienne Nicotra

Field Studies in Ecology 1

Australian National University

eVIEW

Published by ANU eView
The Australian National University
Acton ACT 2601, Australia
Email: enquiries.eview@anu.edu.au
This title is also available online at press.anu.edu.au

National Library of Australia Cataloguing-in-Publication entry

Title: Researching functional ecology in Kosciuszko National Park
            / editors: Hannah Zurcher, Ming-Dao
            Chia, Adrienne Nicotra, Michael
            Whitehead.

ISBN: 9781921934391 (paperback) 9781921934407 (ebook)

Series: Field studies in ecology ; no. 1.

Subjects: Ecology--Research.
            Climatic changes--Australia.
            Plants--Australia.
            Kosciuszko National Park (N.S.W.)

Other Creators/Contributors:
            Zurcher, Hannah, editor.
            Chia, Ming-Dao, editor.
            Nicotra, Adrienne, editor.
            Whitehead, Michael, editor.

Cover design and layout by ANU Press. Cover photograph by Michael Whitehead.

Contents

Editors' note

Thirty undergraduates, 20 research experts, and 10 days in Kosciuszko National Park in the Australian Alps: sounds like a challenge? Well, it was, for everyone involved. For most of the students, this was their introduction to field research. Intensive preparation paired with equally intensive lecture sessions, and students were encouraged to expand on the work of their peers. The course itself was ground-breaking: ANU provides many fantastic field trips, and has a truly extensive range of intensive courses, but nothing quite like BIOL2203 'Field Studies in Functional Ecology'. Students worked in groups of four or five, sharing data and research methods with other groups while investigating challenging problems. Lecturers and 'resource people' supported and mentored students. The format was novel, and immensely successful.

It is from this successful venture that the contents of this inaugural volume of *Field Studies in Ecology* have been sourced. Papers have been selected to represent each experiment conducted. Experiments often had several excellent submissions available, so selecting representative manuscripts was challenging, to say the least. Selected papers were subject to peer review and then revised according to requirements for this publication.

Participants on the course were all keen, intelligent, driven students, high academic achievers in their university life, and all created high-quality research. It is reassuring to know that the future of ecological field research is in such capable hands and minds.

We'd like to thank everyone who has made this volume possible, especially course convener Adrienne Nicotra for her guidance and support. We'd like to thank our submitting authors for the excellent papers, and our reviewers for their constructive comments that improved these papers.

We're grateful to have been given the opportunity to attend this field course, and even more so to have had the opportunity to focus on everyone's hard work and research. We hope that you enjoy reading this volume as much as we enjoyed putting it together!

Hannah Zurcher and Ming-Dao Chia
Student Editors

This volume is the result of dedicated work on the part of my co-editors and all the authors as well as unfailing editorial support from Abigail Widdup, and support to achieve publication from Beth Beckmann, copyeditor Geoff Hunt and staff at ANU Press. The idea of publishing this volume came from the 'Course Books' produced on a field course I attended as a postgraduate student myself. Those books provided a resource for future students in the course, as well as for countless other researchers at the field sites. I hope this one will do the same. In addition, this book will provide a record for the students themselves of their original discoveries and contribution to the field, and, I believe, will be a lasting resource and inspiration.

Adrienne Nicotra
Editor-in-Chief

Syllabus

This is the syllabus that we started with. Weather and other chance events meant we didn't do things exactly when and how we planned. But, we did all of this and more.

time	pre course	Monday 7/12	Tuesday 8/12	Wednesday 9/12	Thursday 10/12	Friday 11/12	Saturday 12/12	Sunday 13/12	Monday 14/12	Tuesday 15/12	Wednesday 16/12	Thursday 17/12	Friday 18/12	Saturday 19/12	post course
7:00		travel to site	breakfast/lunch prep	breakfast/lunch prep	breakfast/lunch prep	breakfast/lunch prep	breakfast/lunch prep	breakfast/lunch prep	breakfast/lunch prep	breakfast/lunch prep	breakfast/lunch prep	breakfast/lunch prep	breakfast/lunch prep	breakfast/lunch prep	
8:00			Before we walk AN/KG/SV	Lecture 1B MW											
9:00	pre course readings and study questions HW	intro talk: ground rules AN/MW/MK	Blue Lake Walk – WK1. Posing questions	Field problem 1A	Field problem 1A	Field problem 1B	Field problem 1B	Break	Field problem 2A	Field problem 2A	Field problem 2B	Field problem 2B	Wrap up projects, finalise data	Coursebook prep and wrap up, data synthesis, metadata submission	Prac write up, final entries in fieldbook
10:00		lunch and settling in							Break						
11:00		lunch and settling in	Wk3: intro. des & data AN		WK5: stats & data AN			Mid course quiz (time tbc)					End of course quiz (fieldbooks)		
12:00		Local natural history walks with staff, resource people and local experts (ALL)			Field problem results followed by talks with resource person	Analysis and interp, prep for presentation	Analysis and interp, prep for presentation	Break		Field problem results followed by talks with resource person		Analysis and interpretation, prep for presentations			
13:00															
14:00			Wk2: bc+f des & data AN					Reflexive disc. BB (fieldbooks)					Spotlight on restoration/ conservation/ management	Personal pack and clean up	
15:00				Wk4: More Collab MW/XW	WK6: graphs & talks SG	Wk7: Even more collab MW/XW	Field problem symposium 1 HW	Lecture 2A AN	Wk8: RIT HW	WK9: Storytelling MW		Field problem symposium 2 SG			
16:00		free time	free time	free time	free time	free time	Break	Lecture 2B MW	free time	free time	free time	Free time	free time	Return to Canberra	
17:00	pre course meeting	dinner	dinner	dinner	dinner	dinner	dinner	dinner	dinner	dinner	dinner	dinner	dinner		
18:00		Lecture 1A AN	Faculty symposium 1 AN	free time	free time	free time	Trivia night	Faculty symposium 2 MW	free time	free time	free time	Free time	Stakeholder dinner/trip slide show		
19:00		Opening night activities	Talks with resource person (tbc)					Talks with resource person (tbc)							
20:00															

ix

Workshop summaries

Scientific research is more complicated a field than most undergraduates imagine it to be, and the usual courses often leave students unprepared for the issues that they will face in their careers as researchers. This course featured workshop sessions covering various aspects of science that are not traditionally elaborated upon in standard degree programs. Workshops touched on a wide range of topics and were taught in an informal fashion.

Workshop 1: Posing questions (self-directed)

Science is a habit as much as a career. Observing one's surroundings and asking questions are habits of scientists, and all questions are important. However, answering is as important as asking; if a scientist asks a question, they should also have created a hypothesis and be thinking of ways to test it. Useful hypotheses are focused and clear, and a null hypothesis is essential. In this workshop, the surrounding environment was important in inspiring questions.

Workshop 2: Experimental design and data handling (Adrienne Nicotra)

Field experiments sometimes offer less control over environmental factors, but good experimental design can go a long way towards ensuring that the data acquired is valid. This workshop showed how, once data is acquired, proper data handling is also useful for maintaining data integrity throughout the analysis process.

Workshop 3: Introduction to collaboration (Xénia Weber and Michael Whitehead)

The vast majority of modern science is intensely collaborative. A successful researcher in today's world will be one who is capable of working well with other scientists. In order to work well with others, a researcher must have a functional understanding of group dynamics, as well as the self-awareness to recognise their failings and their strengths. To that end, this workshop covered standard reflective practices and the basics of self-categorisation.

Workshop 4: More on collaboration (Xénia Weber and Michael Whitehead)

As a continuation of the content covered in Workshop 3, this workshop took knowledge of self-categorisation and moved the emphasis from the self on to the delicate dynamics of the research team. People in established situations often see themselves in specific roles, and the arrival of a new member can be upsetting. This workshop helped students to understand how the newcomer, the established workers and the managers all might respond to make the situation easier.

Workshop 5: Statistical analyses bootcamp (Adrienne Nicotra)

Field research is often seen as largely spending time out in the wild, taking measurements and watching behaviour. However, statistical analysis is one of the more important steps in coming to useful conclusions. In this workshop, different types of data, statistical tests for various types of hypotheses and data visualisation were covered, in the context of exploratory data analysis.

Workshop 6: Presentations (Sonya Geange)

Communication in research is more important than ever. Often, this communication comes in the form of giving presentations. In this workshop, practical elements of presenting were applied through a brief presentation.

Workshop 7: Even more on collaboration (Xénia Weber and Michael Whitehead)

Set late in the course, after students had experienced a range of difficulties in research collaboration, this final teamwork workshop refined the social identity concepts introduced previously. Workshop 7 focused on understanding peers, their behaviour and better ways to ensure that groups worked smoothly together.

Workshop 8: Research integrity (Hannah Windley)

Academic integrity is often presented only in terms of ethics. While that is a significant component, many other issues in research integrity can often affect the quality of research done itself. Keeping good records and using appropriate statistical tests can often affect research conclusions inasmuch as they are important for maintaining integrity. In this workshop, several of the usually less considered topics in research integrity were covered through a series of case studies.

Workshop 9: Scientific storytelling (Michael Whitehead)

The papers published in scientific journals are often all that the world sees of a particular research project. Rarely do they explain the journey to discovery, often vital to the scientific process and giving an incomplete view of the scientific process. In this workshop, the components of a typical scientific paper and the writing of a scientific narrative are discussed.

Biographies of staff and resource people

Core staff

Adrienne Nicotra
biology.anu.edu.au/people/staff-profiles/adrienne-nicotra

Adrienne is a Professor and Convenor of the PhD Program in Evolution, Ecology and Genetics in the ANU Research School of Biology. Adrienne is originally from the United States, and moved to Australia following a PhD researching plant reproductive ecology in the dark, wet understorey of the Costa Rican rainforest. She has been researching plant life in the sunny and dry ever since. Her interests include the adaptive significance of phenotypic plasticity, the evolution of leaf shape, comparative ecology, plant reproductive ecology and plant life in alpine environments. Adrienne's commitment to teaching is shown by her international recognition as a Senior Fellow of the Higher Education Academy.

Michael Whitehead
michaelrwhitehead.wordpress.com

Michael is a Divisional Visitor at the ANU Peakall Lab. He is a plant evolutionary ecologist studying plant interactions above and below ground. He is particularly interested in links between pollinator behaviour and floral evolution, and his research includes population genetics, sensory and behavioural ecology, pollination ecology and phylogenetics. He focuses on organisms often overlooked by other researchers and combines his photographic talents with his research interests to highlight the beauty as well as the complexity of these organisms.

Sonya Geange
sonyageange.com

Sonya is a PhD candidate at the Nicotra Lab, working on a collaborative project conducting a comparative ecological study on phenotypic plasticity in water use traits, using a multisite, multispecies design. She is originally from New Zealand, where her previous work involved investigating aspects of pasture persistence across dairy, sheep and beef systems, and assessing the spread of the invasive Black Beetle (*Heteronychus arator*). Sonya has received recognition as an Associate Fellow of the Higher Education Academy.

Hannah Windley

www.researchgate.net/profile/Hannah_Windley

Hannah completed her PhD in the Foley Lab, working on the role of poisonous plants in the foraging ecology of marsupials. Her work often focuses on the biochemistry of species in remote, dangerous or inaccessible areas. She has travelled widely in the course of her research, and engages actively with the ANU community as well as the international one. As well as being a skilled scientist, she is a talented musician and an able and enthusiastic teacher.

Wes Keys

biology.anu.edu.au/people/wes-keys

Wes provides the staff at the ANU Division of Evolution, Ecology and Genetics with the stellar technical support on which all successful research relies. A jack-of-all-trades and a master of logistics, he supplies and maintains a variety of essential equipment. Wes is the quintessential 'resource person', ensuring that researchers have all the tools they need to do their work and keeps them in wonderful condition. He is also a keen amateur photographer.

Resource people

Nur Abdul Bahar

biology.anu.edu.au/people/nur-abdul-bahar

Nur is a PhD candidate at the ANU Atkin Lab, researching the mechanism underpinning photosynthetic variation in tropical forests. While much of Nur's work takes place in the laboratory rather than in the field, her focus on the impacts of climate change on plant respiration has tremendous impact on future botanical methods. Nur is recognised as an Associate Fellow of the Higher Education Academy.

Phillipa (Pip) Beale

biology.anu.edu.au/people/phillipa-beale

A current PhD candidate at the ANU Foley Lab, Pip is an enthusiastic and cheerful mammal biologist. Beginning her academic career with aspirations towards becoming a veterinarian, Pip decided instead to focus on mammal research. Her current work focuses on aspects of possum chemical biology and metabolism. An artistic soul and able teacher, Pip is a fantastically gifted researcher with astounding communication skills.

Ross Deans

biology.anu.edu.au/people/ross-deans

Ross is a PhD candidate at the ANU Farquhar Lab, working on aspects of stomatal physiology in relation to photosynthetic CO_2 response, utilising both modelling and experimental approaches.

Scott (& Jack) Keogh

biology.anu.edu.au/people/scott-keogh

Professor Scott Keogh completed his BS at the University of Illinois (1991) and MS at Illinois State University (1993) before moving to Australia to do a PhD at the University of Sydney (1993–97). He started an Australian Research Council (ARC) Postdoctoral Fellowship at the University of Sydney in 1997 but moved to a lectureship at ANU in 1998. Since then he has supervised to completion 14 honours students, 3 MPhil students, and 11 PhD students, and hosted 8 ARC Postdoctoral Fellows and multiple postdoctoral research associates. He also served as Convenor of the Graduate Program in Evolution, Ecology and Genetics (2000–11) and as the Associate Director (HDR) for the Research School of Biology (2009–11). He is now serving as the Head of the Division of Evolution, Ecology and Genetics (2012–).

Jack Keogh specialises in lizard-noosing and rock-hopping. He is currently indentured as a part-time laboratory assistant to Scott, performing essential tasks such as entertainment, specimen collection and highly literary lizard-naming. He is presently completing Year 7.

Ulrike Mathesius

biology.anu.edu.au/people/ulrike-mathesius

Ulrike received her Dipl. Biol. (BSc Hons) at the Technical University of Darmstadt in Germany in 1995. She carried out her PhD on the symbiosis between rhizobia and legumes at the ANU Research School of Biological Sciences (RSBS) between 1996 and 1999. This was followed by postdoctoral research at RSBS in the area of plant proteomics between 1999 and 2001. In 2002, she moved to the ANU School of Biochemistry and Molecular Biology with a Postdoctoral Fellowship from the Australian Research Council (ARC). Ulrike then held an ARC Research Fellowship and an ARC Future Fellowship working on the developmental regulation of nodulation, parasitic gall development and lateral root formation in legumes.

Iliana Medina

medinailiana.wix.com/ilianamedina#

Iliana is a PhD candidate at the Langmore Lab, working in brood parasitism in Australian birds. Her work has been well-awarded, receiving commendations from ANU as well as from state and federal government. Her work on avian brood parasitism is highly commended for its conservation implications.

Craig Moritz

biology.anu.edu.au/people/craig-moritz

Professor Craig Moritz did his undergraduate studies at University of Melbourne (1976–79), where he developed his passion for evolutionary biology. For his PhD at ANU (1980–85), he studied chromosome evolution and speciation in arid zone lizards, along the way discovering all-female reproduction in *Heteronotia binoei*. Then he moved across the Pacific Ocean for a postdoc at University of Michigan (1985–88; mitochondrial DNA and evolution of parthenogenesis), before returning to a faculty position at the University of Queensland (1988–2000), including a stint as Head of School. From 2000–12, he was Director of the Museum of Vertebrate Zoology at University of California Berkeley. From mid-2012, he happily settled at the ANU Research School of Biology as a Professor and ARC Laureate Fellow. Craig is also the Director of the joint ANU–CSIRO Centre for Biodiversity Analysis.

Andrew Scafaro

biology.anu.edu.au/people/andrew-scafaro

Andrew is a postdoctoral researcher at the ANU Atkin Lab. He did his PhD at Macquarie University, and currently specialises in plant physiology and molecular biology.

Megan Supple

borevitzlab.anu.edu.au/borevitz-lab-people/megan-supple

Megan is a postdoctoral researcher at the ANU Borevitz Lab, with a PhD in biomathematics from North Carolina State University, with research connecting phenotype, genotype and environment. Her current interests include using genomics tools to understand the evolution of adaptive traits in natural populations, particularly with applications to the conservation of biodiversity. She is currently working on understanding the spatial distribution of neutral and adaptive genomic variation in *Eucalyptus* species, with applications in reforestation appropriate for predicted climate change.

Susanna Venn

susannavenn.wordpress.com/about

Susanna is a postdoctoral plant ecologist with a keen interest in the processes that shape vegetation patterns in alpine areas. She is particularly interested in how snow influences plant community patterns, processes and community (re)assembly. She has previously worked in New Zealand, and has done Australian work for the Global Observation Research Initiative in Alpine Environments (GLORIA, www.gloria.ac.at/). Susanna has been an integral part of the NSW Parks and Wildlife Service Alpine Team, working with stars of Australian alpine research. Most recently, she has been working on shrub, snow and climate feedbacks, hoping to address the plight of alpine areas subject to climate change. Susanna likes to combine field and laboratory work in order to test ecological theory.

Xénia Weber

geogenetics.ku.dk/staff/?pure=en/persons/569542

Xénia completed her honours studies at ANU in 2015 and her skills in education are evidenced by her international recognition as an Associate Fellow of the Higher Education Academy. She commenced her PhD studies in Europe in 2016. She is interested in geography and evolutionary biology, as well as tertiary education.

Supporting cast

Owen Atkin

researchers.anu.edu.au/researchers/atkin-ok

Professor Owen Atkin is the leader of the ANU Atkin Lab, focusing on plant respiration to improve climate change models.

Zak Atkins

www.robertlab.com/Robert_Lab/People.html

Zak is a PhD candidate at La Trobe University whose current research focuses on the Guthega skink, an endangered species found only in the Australian Alps. His work has been central to conservation endeavours. Zak visited us to talk about his work and introduce the skinks.

Elizabeth (Beth) Beckmann

bethbeckmann.wordpress.com

Beth Beckmann is an Australian National Teaching Fellow, Principal Fellow of the Higher Education Academy, Associate Dean in the ANU Science Teaching and Learning Centre, and Reader (Associate Professor) in the ANU Centre for Higher Education, Learning and Teaching. After gaining a degree in biological sciences from the University of Cambridge, Beth studied environmental management and wildlife science in Australia. She became a specialist adviser to national park management agencies on communicating science to the public and influencing visitor behaviour in relation to wildlife. In more recent years, her focus has been university teaching, including the design of courses that focus on research-based learning and the support of university educators, for which she has won several institutional and national awards.

Justin Borevitz

borevitzlab.anu.edu.au/borevitz-lab-people/justin-borevitz-lab-leader

Professor Justin Borevitz is the leader of the ANU Borevitz Lab, working on genomics and phenomics in plants, with particular applications in climate adaptation.

Tim Brown

borevitzlab.anu.edu.au/borevitz-lab-people/tim-brown

Tim is a postdoc researcher in the ANU Borevitz Lab, with a focus on using emerging camera technologies in ecological research. Tim joined the course to demonstrate some of these.

Ken Green

www.aias.org.au/directory/green1.html

Ken is a researcher with the NSW National Parks and Wildlife Service and Global Mountain Biodiversity Assessment committee. Ken spent a day with the course talking about the research he and colleagues have underway in the park.

Stuart Johnston

au.linkedin.com/in/stuart-johnston-074295ba?trk=prof-samename-name

Stuart is the Executive Director of Assets and Network Transformation at Energy Networks Australia, and was previously a Corporate Environmental Manager at TransGrid as well as an Alpine Region Soil Conservation Officer with the NSW National Parks and Wildlife Service. Stuart completed a PhD at ANU specialising in alpine ecology. Stuart participated as a panellist for our 'Spotlight' session.

Mel Schroder

www.researchgate.net/profile/Mellesa_Schroder

Mel is an Environmental Management Officer with the NSW National Parks and Wildlife Service. Her work focuses on the complex task of monitoring the impacts of new and existing recreational infrastructure on the natural environment. Mel joined the course to tell us about research in the park and participated as a panellist for our 'Spotlight' session.

Bindi Vanzella

murrumbidgeelandcare.asn.au/projects/rlf

Bindi is a Riverina Regional Landcare Facilitator, supporting biodiversity conservation, sustainable agriculture and social inclusion. Bindi participated as a panellist for our 'Spotlight' session.

The field problems

Over the course of two weeks each student conducted four group projects. All data were archived and most analyses were conducted as a group during the field trip. Following the field trip, each student selected one of their four projects to write up as a standard research article.

An editorial panel, composed of two student volunteer editors and the course convenors, selected one excellent example of each of the field problems for inclusion in this volume. Those articles were peer reviewed and revised before final acceptance and publication. Any field problem that was not the subject of a full article is included in this volume in summary form.

The full data archive is available to future students on the course web page or by contacting the course convenor.

Investigation of the niche partitioning of selected Ranunculaceae species in Kosciuszko National Park along a soil moisture gradient, by comparison of hydraulic characteristics

Angela Stoddard, Tess Walsh Rossi, Cameron McArthur, Sarah Stock, Ming-Dao Chia, Hannah Zurcher, Chen Liang, Christine Mauger, Julia Hammer

Abstract

Environmental gradients that function as niche partitioning axes underpin plant biodiversity. Alpine zones are recognised biodiversity centres particularly threatened by a changing climate, which may alter such gradients. The soil water content (SWC) of coexisting Ranunculaceae in alpine and subalpine zones of Kosciuszko National Park, Australia (*Caltha introloba, Ranunculus gunnianus, Ranunculus muelleri* and *Ranunculus graniticola*) was investigated as a potential ecological niche division. We hypothesised that study species will be found in soils of distinct SWC, and hydraulic leaf characteristics will vary between species with respect to this gradient. As hypothesised, species were distributed along a statistically distinct soil water gradient ($P<0.001$), with *C. introloba* found in soils of highest SWC. Whilst *R. gunnianus* was found in soils of higher mean SWC than the other *Ranunculus* species studied, the significance of this data could not be determined. In accordance with previous studies, the species of highest SWC, *C. introloba*, was found to have the leaf surface with highest stomatal density ($P<0.0001$); highest adaxial stomatal density ($P<0.0001$); largest xylem diameter; and lowest xylem density within a petiole section. Notably, a distinct trend of higher SLA for species of lower SWC was determined ($P<0.0001$) in contradiction with the literature, which may have novel implications.

Introduction

The coexistence of plant species within a community is commonly attributed to partitioning of resources. These partitions, or ecological niches, often reside along environmental gradients (Silvertown 2004). Niche segregation along such gradients has been described as vital to the success of large, biodiverse plant communities (Silvertown 2004). The characterisation of niche axes has been proposed as a means to model the ecological impacts of a changing environment (Buckley 2013). An observed trend towards homogenisation in ecosystems globally highlights the importance of understanding niche partitioning gradients that underpin plant biodiversity (Clavel et al. 2011).

Alpine environments have been recognised as biodiversity hot spots for flowering plants globally (Hörandl and Emadzade 2011). The Australian alpine region, in particular, has a very small geographical extent, with high endemism and floral diversity. These vegetation communities are under great threat by both climate change and human impacts (Edmonds et al. 2006; Worboys and Good 2011). Conservation of this diversity relies on a developed understanding of the niche systems underpinning the native alpine flora, which encompasses a suite of perennial herb species including the Ranunculaceae family.

Ranunculaceae, or the buttercup family, have shown high adaptive potential to alpine zones on a global scale (Hörandl and Emadzade 2011). Thus, they are a valuable model system relevant to a number of alpine environments worldwide. This study will aim to elucidate some major components of the niche partitioning systems between sympatric Ranunculaceae species within the subalpine and alpine ranges of Kosciuszko National Park. The study species—*Caltha introloba, Ranunculus gunnianus, Ranunculus muelleri* and *Ranunculus graniticola* (Costin et al. 2000)—coexist across a relatively small elevation gradient within alpine herb fields, and therefore are likely to have strong functional niche divisions.

The investigation of niche divisions in plants has been widely achieved by the assessment of functional traits (Buckley 2013). From such studies, soil water content has been identified as a major environmental axis in niche partitioning (Silvertown 2004). Leaf hydraulic traits of *Ranunculus* species have been shown to vary significantly along soil water gradients (Kołodziejek and Michlewska 2015). Such traits include adaxial and abaxial stomatal density, xylem diameter, xylem density and specific leaf area (SLA).

In previous studies of Ranunculaceae species, stomatal density on both adaxial and abaxial surfaces has been reported to have a strong positive correlation with soil water content (SWC) (Lynn and Waldren 2002; Kołodziejek and Michlewska 2015). This correlation reflects an adaptive trade-off between water loss through evaporative transportation and influx of carbon dioxide for photosynthesis. A study of *Ranunculus repens* reported that individuals in wet soils had higher relative adaxial density (Lynn and Waldren 2002). This trend is consistently observed as water plants do not require stomata on the underside of submerged leaves, and are not limited by water loss from evaporative transpiration on adaxial surfaces (Xu and Zhou 2008).

In studies of the vascular characteristics of flowering plants under various soil moisture regimes, high SWC was strongly associated with larger mean xylem diameter and reduced xylem tissue as a percentage of petiole area (Awad et al. 2010). A large xylem diameter increases the efficiency of water conductivity in conjunction with embolism risk (Awad et al. 2010). As plants in high SWC have an exceptionally low embolism risk, wider xylem provides an adaptive advantage. It follows that increased water transport efficiency is associated with reduced investment in net xylem tissue.

Furthermore, high SLA is often associated with plants in high SWC, featuring leaves with high growth rates, high photosynthetic rates, a low carbon investment and relatively short lifespan. Conversely, a low SLA is associated with tough, long-lived leaves that are thick and/or have high dry matter content (Pérez-Harguindeguy et al. 2013). In previous studies of *Ranunculus*, thicker leaves with thicker epidermis were observed at drier test sites, typical of a low SLA (Cunningham et al. 1999; Kołodziejek and Michlewska 2015).

It has been previously hypothesised that the functional niches of the Ranunculaceae species studied so far are primarily partitioned by inhabiting soils of varied SWC (Armstrong 2003). Thus, this study will foremost aim to test the hypothesis that (1) *R. graniticola*, *R. muelleri*, *R. gunnianus* and *C. introloba* will be found in soils of significantly different SWC. Based on previous reports (Costin et al. 2000) we also predict that (2) *C. introloba* will inhabit areas of the highest SWC and (3) *R. gunnianus* may be characterised by soils of slightly higher SWC than *R. graniticola* and *R. muelleri*.

We will investigate the aforementioned hydraulic characteristics of each species, and analyse them with respect to the SWC data collected. Based on trends observed in the literature, we hypothesise that species in areas of higher SWC will display (4) higher stomatal density, (5) higher relative adaxial stomatal density, (6) larger mean xylem diameter, (7) lower percentage xylem tissue per petiole area and (8) higher SLA.

Methods

The distribution of Ranunculaceae was observed along a transect from Charlotte Pass to Blue Lake (Table 1). Patches separated by at least 10 m were considered individual sites, from which one individual plant was sampled. Soil water content was measured using the Delta T Theta Probe, which was inserted into the soil at three points in an approximately triangular pattern around the plant to generate an average value.

Table 1: Local distribution of Ranunculaceae study species within Kosciuszko National Park.

Species	Vegetation community	Sample size (n)
Caltha introloba	Short alpine herbfield: often found below snow patches and in snow melt streams.	6
Ranunculus gunnianus	Damp areas within tall alpine herbfield and sod tussock grassland	15
Ranunculus muelleri	Tall alpine herbfield and sod tussock grasslands	23
Ranunculus graniticola	Tall alpine herbfield and sod tussock grasslands	18

The number of individuals of each species included in this analysis is given as the sample size.

Source: Adapted from Costin et al. (2000).

At each site, one sample plant was dug up minimising root mass disturbance, wrapped in damp paper towel and sealed in a plastic zip-lock bag. At least five samples of each species were collected from individual sites and stored in a cooler room at approximately 4°C. These samples were utilised in further lab analysis of hydraulic characteristics: stomatal density, xylem diameter, xylem density and SLA.

Stomatal density

Stomatal density was defined as the mean total area of stomata for a given area of leaf surface, given as a percentage. To measure this, leaves were randomly selected from collected samples (n=2) of each species to generate stomatal peels. Leaves were pressed into a 1 mm film of super glue on a glass microscope slide, one abaxial and one adaxial side down. Once glue was set, leaves were removed leaving a negative impression. Slides were viewed and photographed under a light microscope at 100× magnification. Using the software ImageJ, the percentage area of stomata per leaf area was approximated by selecting random sample areas of 1 mm². Random sampling of stomatal was performed twice per stomatal peel, resulting in a sample size of (n=4) per species.

Vascular profile

Petiole sections were created from samples of each species (n=2) by slicing the hydrated petioles in a solid, stabilising medium with a razor. Sections were kept immersed in water, and dyed in a dilute Toluidine blue solution. Sections were transferred to a glass slide, covered with a cover slip and viewed under a light microscope. Petioles were photographed at 100× magnification, and vascular bundles at 400× magnification. ImageJ was used to measure the mean diameter of xylem vessels within samples from 400× images. Similarly, ImageJ was used to measure both the total area of the petiole section and the xylem tissue from 100× images. This data was expressed as xylem tissue as a percentage area of the petiole.

Specific leaf area

Specific leaf area (SLA) refers to the one-sided area of a hydrated leaf divided by its dry mass, measured in cm²/g. The one-sided surface area of randomly selected leaves from samples (n=5; n=4 for *R. gunnianus*) was measured using ImageJ to analyse scanned images. Leaves were then dehydrated by microwaving them between two pieces of cardboard on low-medium power for 2 minutes. Dry mass was determined using an electronic balance.

Statistics

Results were analysed using a two-way ANOVA test. Means with $P<0.05$ were considered significantly different.

Results

The mean SWC of each Ranunculaceae sample site was found to vary significantly as predicted by hypothesis (1) (*P*<0.001) (Figure 1). The species occupying the sites of highest soil moisture was *C. introloba* as predicted (2) with a mean of 71.7 per cent, whilst *R. graniticola* sites had the lowest SWC of 19.7 per cent. There was, however, no significant difference evident between *R. gunnianus* and *R. muelleri* in SWC content, of 35.8 per cent and 30.7 per cent respectively. This data neither supports nor conflicts with hypothesis (3).

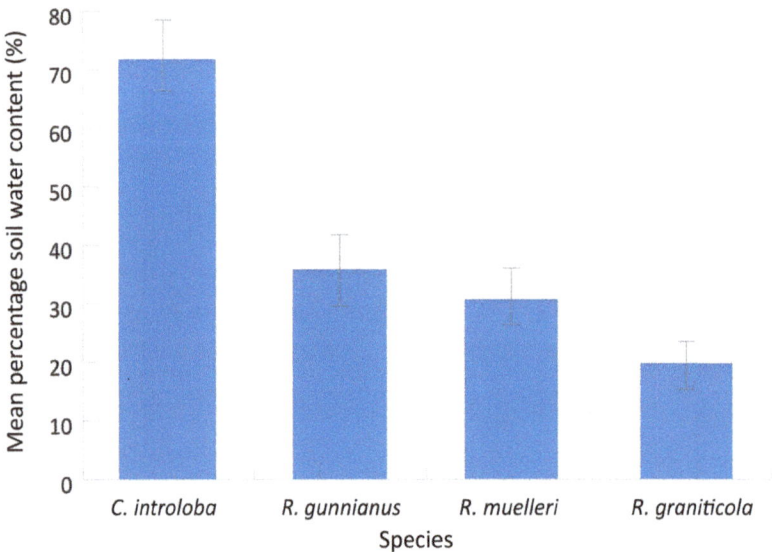

Figure 1: Mean percentage soil water content of *Caltha introloba* (n=6), *Ranunculus gunnianus* (n=15), *Ranunculus muelleri* (n=23) and *Ranunculus graniticola* (n=18).

Note: Error bars are ±SE. Using ANOVA, *P*<0.001.

Source: Results generated from field data using a soil moisture probe.

Mean stomatal density was not shown to have any significant or consistent correlation to soil moisture gradient as was anticipated (4) (Figures 2 and 3). However, the mean stomatal density was found to be significantly different between species on both abaxial (*P*<0.0001) and adaxial (*P*<0.0001) surfaces. As hypothesised (4), *C. introloba* displayed the highest stomatal density, 4.1 per cent. Notably, *C. introloba* stomatal density was exclusively found on the adaxial surface, which was not in conflict with expectations (5).

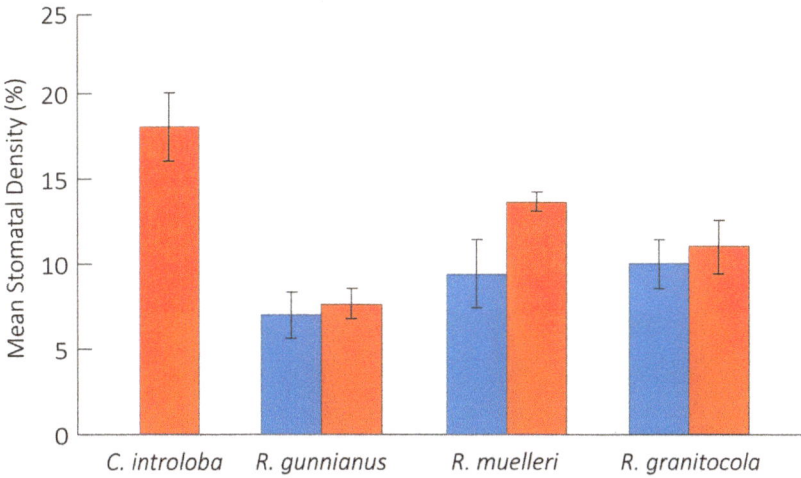

Figure 2: Mean stomatal density of the abaxial (blue) and adaxial (red) leaf surfaces of *Caltha introloba*, *Ranunculus gunnianus*, *Ranunculus muelleri* and *Ranunculus graniticola*. Sample size of n=4 in all species.

Note: Error bars are ±SE. Using ANOVA, mean adaxial ($P<0.0001$) and abaxial ($P<0.0001$) stomatal densities were significant.

Source: Results generated by analysis of light microscope images of stomatal peels at 100X using ImageJ.

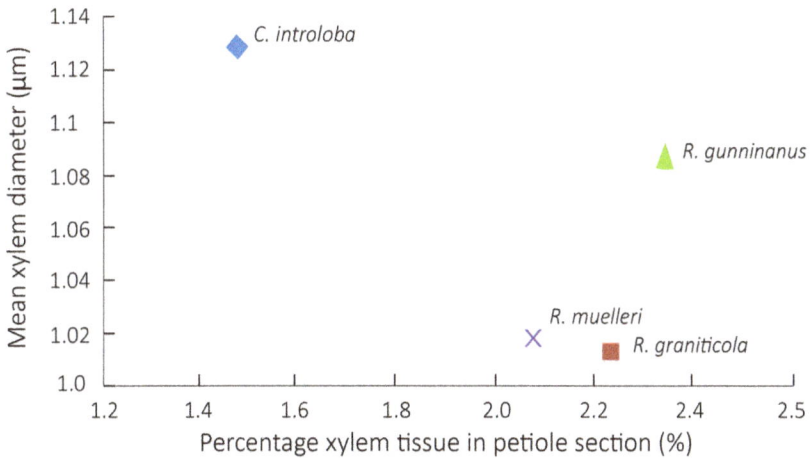

Figure 3: Mean diameter of xylem vessels and the percentage area of xylem tissue in a petiole section for *Caltha introloba*, *Ranunculus gunnianus*, *Ranunculus muelleri* and *Ranunculus graniticola*.

Note: Sample size of n=2 in all species.

Source: Results generated by analysis of light microscope images of petiole sections at 100X and 400X using ImageJ.

The percentage xylem tissue in the total area of a petiole section and the mean diameter of xylem vessels supported some predicted trends (Figures 4 and 5). As hypothesised (6), *C. introloba* had the largest mean xylem diameter of 1.13 μm, and the lowest mean percentage of xylem tissue in the petiole section of 1.5 per cent (7). Whilst *R. gunnianus* had a similar percentage xylem tissue in the petiole as the other *Ranunculus* species, it had a notably larger mean xylem diameter. *R. muelleri* and *R. graniticola* did not vary significantly in either trait.

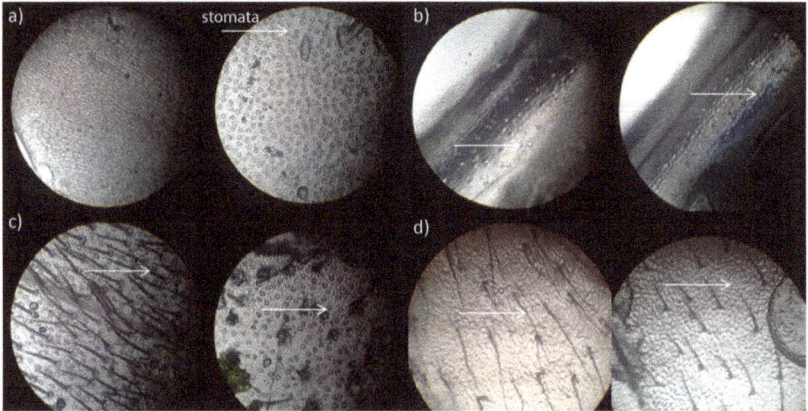

Figure 4: Images from light microscope at 100X magnification. Slides depict impressions of stomatal peels prepared using superglue adhesive. Left image in each pair is abaxial side, right image is adaxial: a) *Caltha introloba*, b) *Ranunculus gunnianus*, c) *Ranunculus muelleri* and d) *Ranunculus graniticola*.

Source: Authors' photograph.

SLA did not follow a positive trend with SWC as predicted (8) (Figure 6). In contradiction, *R. graniticola*, the species with lowest SWC, had a significantly larger SLA of 146 cm^2/g than all other species ($P<0.0001$). Overall, no consistent trend was observed along a soil moisture gradient.

Discussion

This investigation yielded ample evidence in support of the proposal of a soil moisture gradient comprising a functional niche division for the sympatric species *Caltha introloba*, *Ranunculus gunnianus*, *Ranunculus muelleri*, and *Ranunculus graniticola* in Kosciuszko National Park. Through the investigation of hydraulic characteristics, some correlations were observed in stomatal density and vascular composition with respect to this gradient as predicted. However, a potentially novel relationship was observed between SWC and SLA.

Soil moisture gradient

SWC varied significantly between the study species (1) (Figure 1), providing strong evidence in favour of SWC as a niche partitioning axis. As predicted (2), *C. introloba* was found in soils of the highest SWC. Whilst *R. gunnianus* was found in soils of slightly higher mean SWC than *R. muelleri* and *R. graniticola* (3), further investigation is required to substantiate this relationship. Notably, all observations were consistent with the literature (Costin et al. 2000). As soil water is arguably one of the most influential abiotic factors on plant life strategies (Silvertown 2004; Kołodziejek and Michlewska 2015), we expect the distribution of the species along this axis to also reflect trends in intrinsic hydraulic characteristics.

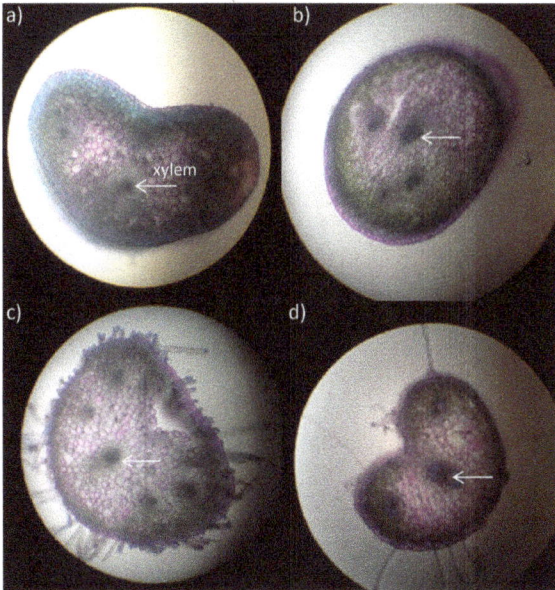

Figure 5: Images from light microscope at 100X magnification. Slides depict hydrated petiole sections, dyed using Toluidine Blue. As labels indicate, one example image is provided for each species studies: a) *Caltha introloba*, b) *Ranunculus gunnianus*, c) *Ranunculus muelleri* and d) *Ranunculus graniticola*.

Source: Authors' photograph.

As the study species were found to be distributed along a SWC gradient, analysis of trends in hydraulic characteristics with respect to this gradient is assumed to be valid. We do note that data collection using the soil moisture probe was subject to human bias in selection of trial points. Sample sites and plants within sites were also selected with a slight bias in favour of visibility and accessibility. In general, this was a consequence of the large transect used, the use of artificial walkways, the limited access to sites in rough terrain and sites obscured by other landmarks or flora. This bias may have resulted in the preferential sampling of species from cleared areas or drier soils near the bare walkways.

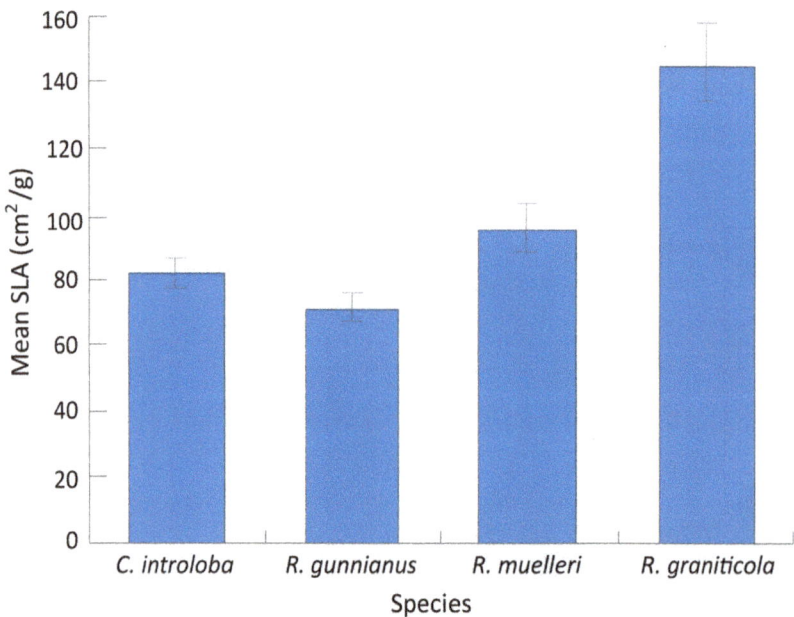

Figure 6: Graphical comparison of the mean SLA of *Caltha introloba* (n=5), *Ranunculus gunnianus* (n=4), *Ranunculus muelleri* (n=5) and *Ranunculus graniticola* (n=5).

Note: Error bars are ±SE. Using ANOVA, mean values were found to be significant ($P<0.0001$).

Source: Authors' data.

Stomatal density

The highest mean stomatal density was observed in *C. introloba*, the species inhabiting the wettest soils (4) (Figures 2 and 3). As stomatal density positively correlates to gas exchange and carbon assimilation (Xu and Zhou 2008), we propose that *C. introloba* has adapted to capitalise on these traits given the high water availability. While this finding is in agreement with previous studies (Lynn and Waldren 2002; Kołodziejek and Michlewska 2015), no universal trend in stomatal density with respect to SWC was apparent in the data.

Furthermore, the adaxial stomatal density of *C. introloba* was significantly greater than all *Ranunculus* species as hypothesised (5) (Lynn and Waldren 2002). We also note that *C. introloba* stomatal density was exclusive to the adaxial surface, which is an adaptation common to plants partially submerged in water (Kirkham 2004).

We recognise potential sources of error in the microscope image analysis methodology. Images were taken using a hand-held camera to the eyepiece rather than an attached camera. The ImageJ software was calibrated once per magnification, whilst an accurate analysis of area and length would demand minor recalibrations for each image.

Vascular profile

Wider xylem and less percentage xylem tissue in a petiole section were observed for *C. introloba*, the species found in soils of highest SWC, as predicted (6)(7) (Figures 4 and 5). In accordance with SWC data, *R. gunnianus* also had notably wider xylem vessels than *R. graniticola* and *R. muelleri*. Whilst our data aligned with expectations, no significant trend could be established between vascular profile and mean SWC due to small sample sizes. We note that potential sources of error in microscope image analysis are relevant here, as previously described.

However, we recognise the biological significance of the trend between plant vascular profile and SWC that was observed. Wider xylem provides a functional advantage in water conduction efficiency (Awad et al. 2010). In high SWC, risk of xylem cavitation remains low, so adaptive trade-off tend to favour wide vessels (Awad et al. 2010). Likewise, reduced xylem tissue observed in *C. introloba* would conserve energy, as less vasculature is required to transport the equivalent water in high SWC.

Given the distinct vascular profile of *C. introloba* (Figure 4), the drying of soils and the disruption of water-flow patterns in the Australian Alps (Edmonds et al. 2006; Worboys and Good 2011) may have implications for species success in the future. As evidence indicates that *C. introloba* inhabit a niche partition at the high end of a soil moisture gradient, specialist vascular characteristics may endanger long-term species success within the landscape (Clavel et al. 2011). In particular, reduced SWC may result in higher rates of fatal embolism of the plant water column, which is associated with wide xylem. Trait plasticity experiments would be essential in predicting the extent of this effect.

Specific leaf area

SLA was significantly higher in study species found in drier soils, which directly refuted our hypotheses (8) (Figure 6). The highest SLA was recorded for *R. graniticola*, the species inhabiting the lowest SWC. The two lowest SLA values were recorded for *C. introloba* and *R. gunnianus*, which occupy the high end of the proposed soil moisture gradient. These results contradict previous studies that report that low SLA, and therefore greater long-term investment in leaf structure, is more strongly selected for in low water resource environments (Cunningham et al. 1999; Kołodziejek and Michlewska 2015).

A similarly unusual case of low SWC correlating with high SLA was observed in a study of silver birch trees, where individuals exposed to drought displayed an increase in SLA (Aspelmeier and Leuschner 2006). It was suggested that production of thinner, less costly leaves in drought conditions was the cause of high SLA. However, a high SLA in low SWC remains a highly unusual result for plant species (Pérez-Harguindeguy et al. 2013), which may be due to inadequacies of the experimental method.

A potential reason for the novel results generated is related to the atypical leaf drying method (Pérez-Harguindeguy et al. 2013). In the case of partial dehydration, species from high SWC like *C. introloba* may have retained more water, leading to the lower SLA observed. Alternatively, pronounced differences in growth and/or flowering times (Pickering 1995) may have resulted in smaller leaves in species from high SWC, again skewing results. Further experimentation is required to determine if this novel result is valid, which may suggest interesting biological implications.

Future directions

Future direction may involve both refining and broadening the scope of this experiment. To refine, larger sample sizes using standardised techniques and units would greatly enhance the quality and comparability of data generated. In determining soil water content, a more robust sampling regime is recommended for future investigation. For further studies of stomatal density and vascular data, use of a mounted camera on the light microscope is suggested to ensure accurate image analysis. Additionally, the expression of stomatal density in number/mm^2, a widely utilised unit, would increase comparability with previous studies (Kołodziejek and Michlewska 2015). We also recommended that leaves be oven-dried to determine dry mass for SLA calculations as outlined in the literature (Pérez-Harguindeguy et al. 2013).

Furthermore, a broader scope of study may investigate other hydraulic characteristics, such as xylem conductivity, leaf water potential, vulnerability to embolism, root system morphology (Pérez-Harguindeguy et al. 2013), pubescence, epidermal thickness, relative conductivity of adaxial and abaxial stomata (Kołodziejek and Michlewska 2015) and water use efficiency (Xu and Zhou 2008).

In addition to this, investigation of the plasticity of the hydraulic traits studied would be a valuable addition to this area of study. Our experimental design relies on the assumption that the trends in phenotype are predominantly genotypic, or that plants only successfully inhabit soils of approximately the same SWC in which they were observed. Examining the plasticity of these traits in response to varied SWC in a greenhouse setting would allow a more refined understanding of the strength of SWC as a niche division, as well as an enhanced ability to make predictions about the fate of Ranunculaceae in the Australian Alps.

In addition, more species of Ranunculaceae within the same distribution may be investigated, in the context of appropriate permits: *R. anemoneus*, *R. millani*, *R.dissectifolius* and *R. niphophilus* (Costin et al. 2000). This would support a much more comprehensive profile of Ranunculaceae niche partitioning in the Kosciuszko alpine and subalpine zone.

Conclusions

We found substantial evidence of niche partitioning of *C. introloba*, *R. gunnianus*, *R. muelleri* and *R. graniticola* along a soil water gradient (1), with *C. introloba* found in soils of highest SWC as predicted (2). However, found only weak evidence that *R. gunnianus* inhabits a niche of higher SWC than *R. muelleri* and *R. graniticola* (3).

Through comparative analysis, hydraulic traits were found to correlate to relative SWC in some cases. The species found in highest SWC, *C. introloba*, was found to have the leaf surface with highest stomatal density (4) and highest relative adaxial stomatal density (5) as predicted. However, this trend was not observed across all four species. Likewise, *C. introloba* was observed to have wider xylem (6) and a low relative density of vascular tissue (7) within a petiole section as hypothesised. *R. gunnianus*, found to occupy a slightly higher position along the SWC gradient, was also observed to have wider xylem than other *Ranunculus* species. However, vascular observations were limited due to the small sample size. Notably, a distinct trend of higher SLA for species of lower SWC was determined in contradiction with the literature (8), which may have novel implications.

Acknowledgements

Thanks to the Research School of Biology and The Australian National University for providing all of the equipment and support to make this research possible. Thank you to NSW National Parks and Wildlife Services for allowing us to undertake this field study within Kosciuszko National Park. Lastly, thank you to the Southern Alps Ski Lodge for their generosity in hosting the group. We thank the resource people for their guidance.

References

Armstrong T (2003) Hybridization and adaptive radiation in Australian alpine *Ranunculus*. PhD thesis, The Australian National University.

Aspelmeier S, Leuschner C (2006) Genotypic variation in drought response of silver birch (*Betula pendula* Roth): Leaf and root morphology and carbon partitioning. *Trees – Structure and Function* 20: 42–52. doi. org/10.1007/s00468-005-0011-9

Awad H, Barigah T, Badel E, Cochard H, Herbette S (2010) Poplar vulnerability to xylem cavitation acclimates to drier soil conditions. *Physiologia Plantarum* 139: 280–8. doi.org/10.1111/j.1399-3054.2010.01367.x

Buckley LB (2013) Get real: Putting models of climate change and species interactions in practice. *Annals of the New York Academy of Sciences* 1297: 126–38. doi.org/10.1111/nyas.12175

Clavel J, Julliard R, Devictor V (2011) Worldwide decline of specialist species: Toward a global functional homogenization? *Frontiers in Ecology and the Environment* 9: 222–8. doi.org/10.1890/080216

Costin AB, Gray M, Totterdell CJ, Wimbush DJ (2000) *Kosciuszko Alpine Flora*, 2nd edn. CSIRO Publishing, Melbourne.

Cunningham S, Summerhayes B, Westoby M (1999) Evolutionary divergences in leaf structure and chemistry, comparing rainfall and soil nutrient gradients. *Ecological Monographs* 69: 569–88. doi. org/10.1890/0012-9615(1999)069[0569:EDILSA]2.0.CO;2

Edmonds T, Lunt I, Roshier D, Louis J (2006) Annual variation in the distribution of summer snowdrifts in the Kosciuszko alpine area, Australia, and its effect on the composition and structure of alpine vegetation. *Austral Ecology* 31: 837. doi.org/10.1111/j.1442-9993.2006.01642.x

Hörandl E, Emadzade K (2011) The evolution and biogeography of alpine species in *Ranunculus* (Ranunculaceae): A global comparison. *Taxon* 60: 415–26.

Kirkham MB (2004) *Principles of Soil and Plant Water Relations*. Elsevier Academic Press, USA.

Kołodziejek J, Michlewska S (2015) Effect of soil moisture on morpho-anatomical leaf traits of *Ranunculus acris* (Ranunculaceae). *Polish Journal of Ecology* 63: 400–13. doi.org/10.3161/15052249PJE2015.63.3.010

Lynn DE, Waldren S (2002) Physiological variation in populations of *Ranunculus repens* L. (Creeping buttercup) from the temporary limestone lakes (turloughs) in the west of Ireland. *Annals of Botany* 89: 707–14. doi.org/10.1093/aob/mcf125

Pérez-Harguindeguy N, Díaz S, Garnier E, Lavorel S, Poorter H, Jaureguiberry P, Bret-Harte MS, Cornwell WK, Craine JM, Gurvich DE, Urcelay C, Veneklaas EJ, Reich PB, Poorter L, Wright IJ, Ray P, Enrico L, Pausas JG, de Vos AC, Buchmann N, Funes G, Quétier F, Hodgson JG, Thompson K, Morgan HD, ter Steege H, van der

Heijden MGA, Sack L, Blonder B, Poschlod P, Vaieretti MV, Conti G, Staver AC, Aquino S, Cornelissen JHC (2013) New handbook for standardised measurement of plant functional traits worldwide. *Australian Journal of Botany* 61: 167–234. doi.org/10.1071/BT12225

Pickering CM (1995) Variation in flowering parameters within and among five species of Australian alpine *Ranunculus*. *Australian Journal of Botany* 43: 103–12. doi.org/10.1071/BT9950103

Silvertown J (2004) Plant coexistence and the niche. *Trends in Ecology and Evolution* 19: 605–11. doi.org/10.1016/j.tree.2004.09.003

Worboys GL, Good RB (2011) *Caring for our Australian Alps Catchments: Summary Report for Policy Makers*. Department of Climate Change and Energy Efficiency, Canberra.

Xu Z, Zhou G (2008) Responses of leaf stomatal density to water status and its relationship with photosynthesis in a grass. *Journal of Experimental Botany* 59: 3317–25. doi.org/10.1093/jxb/ern185

Analysing phenotypic variation in *Eucalyptus pauciflora* across an elevation gradient in the Australian Alps

Giles Young, Islay Andrew, Kristi Lee, Xiaoyn Li, Rachael Robb, Isabella Robinson, Holly Sargent, Bronte Sinclair

Abstract

Research related to the phenotypic plasticity of species that face rapidly changing conditions in the near future, such as those in the Australian Alps, is extremely important for ecological conservation efforts. *Eucalyptus pauciflora* is a species found throughout much of the Australian Alps. In this paper, the phenotypic plasticity exhibited by the species in terms of tree height and leaves across an elevation gradient was studied to gain insight into how the species is able to survive in such a range of conditions. Height and leaf size decreased with elevation, chlorophyll content (a measure of photosynthetic potential) increased, while specific leaf area (indicative of investment in photosynthesis and growth) and leaf dry matter content (indicative of investment in structural strength) showed no significant trends across the elevation range. These results give insight into the phenotypic plasticity of *E. pauciflora*, and provide information on how the ecosystem may respond to climate change in the future.

Introduction

All around the world, plant species exhibit phenotypic variation across their ranges, because varying environmental conditions cause either the formation of hybrid zones through natural selection or variation within a species (Holman et al. 2011). Intraspecific variation may be due to phenotypic plasticity, namely the ability of a certain genotype to express a variety of phenotypes in response to environmental conditions (Andrew et al. 2010).

This phenotypic change due to the environment leads to measurable correlations between the phenotype of plants and their microhabitat. Understanding these correlations enables more accurate predictions of how ecosystems will respond to climate change, providing information on the resilience of an ecosystem. This allows different ecosystems to be assessed for their vulnerability to climate change.

Alpine regions are seen to undergo a more rapid response to climate change than is seen in other areas, making them a priority in conservation efforts (Jurasinski and Kreyling 2007). These areas have variable and highly localised abiotic conditions, which are accompanied by high biodiversity and relatively sharp transitions between vegetation communities (Beniston 2003). Due to the complex nature of these ecosystems, change in the environment is thought to cause homogenisation of high altitude communities and loss of alpine biodiversity (Speed et al. 2012).

It is therefore important to explore the ways through which species are able to survive in variable conditions. In this study in the Australian Alps, *Eucalyptus pauciflora*, commonly known as snow gum, was chosen due to its abundance across a large altitudinal range and thereby across a range of abiotic conditions (O'Sullivan et al. 2013).

In order to explore the ways through which *E. pauciflora* is able to survive across different environments, the phenotypic variations in tree height, leaf size, leaf dry matter content (LDMC), leaf chlorophyll content and specific leaf area (SLA) of snow gum trees were measured across an altitudinal gradient. Due to increasing environmental disturbance, tree height and leaf size are usually seen to decrease with elevation (Wang et al. 2012; Royer et al. 2008). LDMC is related to the structural strength of the leaves, while SLA is related to potential photosynthetic and growth rates and therefore an inverse relationship with LDMC is often seen (Pescador et al. 2015). Chlorophyll content, the primary light-harvesting pigment in a plant's leaves, which varies in response to light and stress levels, is vitally important as photosynthetic potential and primary production is limited by the solar radiation absorbed by the plant (Turkis and Ozbucak 2010). In terms of chlorophyll content variability across an elevation gradient, conflicting results have previously been found (Filella and Peñuelas 1999; Covington 1975).

The aim of this paper is to explore the phenotypic variation among *E. pauciflora* across an elevational gradient in the Australian Alps to gain insight into the species' ability to survive in a range of environmental conditions. Leaves are expected to be smaller, LDMC lower and SLA higher at lower elevations, while trees are expected to be shorter at higher elevations.

Method

Field procedure

Sampling was done at four different sites along an elevational gradient (shown in Table 1). At each site, eight leaves were taken from each of 10 different trees, and chlorophyll content (in SPAD units) of three of the leaves from each tree was measured three times with a chlorophyll meter and the measurements from each tree were then averaged. The height of each tree was calculated using angles measured with a clinometer after walking 10 m from the base of the tree. The global positioning system (GPS) coordinates of each tree were also recorded.

Table 1: Elevations of the four different sample sites

Site	Name	Elevation (m)
1	Waste Point	920
2	Kosciuszko Learning Centre	1,200
3	Rainbow Lake	1,600
4	Charlotte Pass	1,860

Source: Authors' data.

Laboratory procedure

In the laboratory, five leaves from each tree were scanned using ImageJ to give an average leaf area for each site, with the three remaining leaves being set aside for later genetic testing. The scanned leaves were then saturated with water by soaking for a minimum of 10 minutes, at which point they were measured using a fine scale balance to determine saturated leaf mass. The leaves were then placed in an oven overnight at 70–80°C, after which they were again weighed to determine oven-dry leaf mass. These measurements were then used to calculate SLA and LDMC using the equations shown below.

$$SLA = \frac{Leaf\ Area\ (mm^2)}{Leaf\ Dry\ Mass\ (g)}$$

$$LDMC = \frac{Leaf\ Dry\ Mass\ (g)}{Saturated\ Dry\ Mass\ (g)}$$

Results

Trees at site 1 were significantly taller than those at Sites 2 and 3 ($P<0.05$, Figure 1a), and while trees at site 1 appeared to be taller than those at site 4, this difference was not significant ($P>0.05$, Figure 1a), due to the large variance seen at site 4. Sites 2, 3 and 4 were also not significantly different from each other ($P>0.05$, Figure 1a).

In terms of leaf size, leaves were seen to decrease in size as elevation increased. Leaf area decreased as elevation increased, though the difference between the leaf size at sites 1 and 2 was not statistically significant ($P>0.05$, Figure 1b). Despite this, the decrease in leaf size from sites 2 to 3 and from sites 3 to 4 was seen to be statistically significant ($P<0.05$, Figure 1b). Leaf mass also appeared to decrease as elevation increased, and again the decrease in mass from sites 1 to 2 was not statistically significant ($P>0.05$, Figure 1c). The decrease in mass from sites 2 to 3 and from 3 to 4 though was statistically significant ($P<0.05$, Figure 1c). No significant trend was seen in regards to SLA ($P>0.05$, Figure 1f).

In the analysis of leaf composition, no significant trend was seen in LDMC across the elevation gradient ($P>0.05$, Figure 1d), but a trend was seen in chlorophyll content. Sites 1 and 2 were seen to have significantly lower chlorophyll content than sites 3 and 4 ($P<0.05$, Figure 1e). However, the difference in chlorophyll content between sites 1 and 2 and between sites 3 and 4 was not statistically significant ($P>0.05$, Figure 1e). There were no significant differences in SLA between sites (Figure 1f).

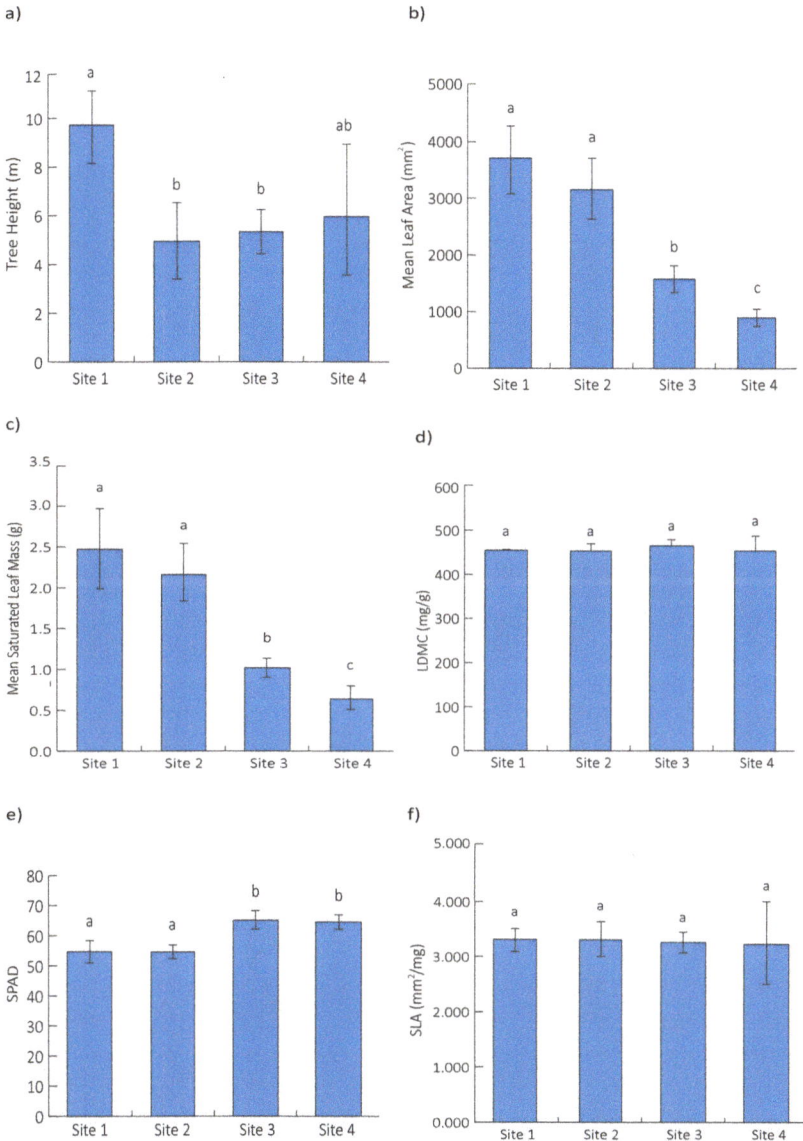

Figure 1: Column graphs showing variation in a) tree height, b) mean leaf area, c) mean saturated leaf mass, d) leaf dry matter content, e) chlorophyll content and f) specific leaf area across an elevational gradient. Site 1 was Waste Point (920 m), site 2 was Kosciuszko Learning Centre (1,200 m), site 3 was Rainbow Lake (1,600 m) and site 4 was Charlotte Pass (1,860 m).

Note: Error bars represent 95 per cent confidence intervals.

Source: Authors' data.

Discussion

In order to better understand how *E. pauciflora* is able to survive in a range of environmental conditions, a number of plant traits were measured across an elevational gradient in the Australian Alps. As seen in Figure 1a, site 1 (the lowest elevation) had taller trees than higher elevation sites. At sites 2 and 3, sampling was limited to small trees due the fact that the leaves of tall trees (which were more abundant) were inaccessible given the equipment available. If this sampling issue could be resolved, it is expected that, given both the hypothesis and results seen previously in the literature (Wang et al. 2012), observations would show tree height decreasing as elevation increases,

In terms of leaf size, when analysed in regards to both leaf area and leaf mass (Figures 1b and 1c), a downward trend is apparent as elevation increased. This is consistent with our predictions and results seen in previous research (Royer et al. 2008).

The results for LDMC and SLA (Figures 1d and 1f) showed no significant trend across the elevational gradient. While these results do not support the hypothesis, previous studies have revealed varying trends in SLA and LDMC across elevational gradients (Zhong et al. 2014). The reason for this deviance from the hypothesis could be due to the genotype of *E. pauciflora* not allowing for significant phenotypic plasticity in regards to SLA or LDMC.

Chlorophyll content was also seen to change with elevation, with the leaves from sites 3 and 4 having significantly higher chlorophyll content than those from sites 1 and 2. In previous studies, a variety of trends have been seen (Filella and Peñuelas 1999; Covington 1975). A possible explanation for the trend seen in this study is that, in order to maintain carbon uptake as leaf size decreases with elevation, an increased chlorophyll content is needed to increase the amount the overall solar radiation absorbed. Another explanation could be that chlorophyll content varies seasonally and in growth periods (Turkis and Ozbucak 2010), such that the observed difference in chlorophyll content may simply be due to differences in growth periods brought on by the varying microclimates at each site.

There is much room for further research into the phenotypic plasticity of *E. pauciflora*, especially in regards to LDMC, SLA and chlorophyll content. In the case of chlorophyll content, measuring it at various periods throughout the year in order to isolate the variable of growth periods may help to increase the accuracy of the results achieved. For SLA and LDMC, analysis of other species within the range may help to determine whether the absence of a clear trend in *E. pauciflora* was due to its genotype or simply the environment in which it is found. Research on additional plant traits could also be done. Additionally, in the analysis of the results achieved for chlorophyll content, leaf area and leaf mass, a clear break was seen between the observations of site 2 and 3 (see Figures 1b, 1c, 1e). Further sampling between sites 2 and 3 may determine whether there is a particular elevation at which traits change, or whether the significant difference in observations was due to the large distance between sites (both elevation and horizontal distance). For some measurements, namely tree height, LDMC and SLA, site 4 exhibited large variation. This may be due to there being more varied microhabitats present at such a high and exposed elevation. Future studies could also investigate the difference in microhabitats in these types of environments. Finally, the leaves set aside for later genetic testing could also provide definitive evidence as to whether the variation in phenotype is due to phenotypic plasticity or genetic variation.

The results achieved in this paper provide insight into the ability of *E. pauciflora* to thrive across such a range of environments. By learning which adaptions enable it to perform better in certain environments, we can better predict how the distribution and characteristics of the snow gum population will change as the climate warms. Through further research, both into *E. pauciflora* and other species, a better understanding can be gained of how the alpine ecosystem as a whole will respond to climate change.

Acknowledgements

We thank The Australian National University Research School of Biology for providing the materials needed to perform the experiment and Megan Supple for supervising the study.

References

Andrew R, Wallis I, Harwood C, Foley W (2010) Genetic and environmental contributions to variation and population divergence in a broad-spectrum foliar defence of *Eucalyptus tricarpa*. *Annals of Botany* 105: 707–17. doi.org/10.1093/aob/mcq034

Beniston M (2003) Climatic change in mountain regions: A review of possible impacts. *Climatic Change* 59: 5–31. doi.org/10.1007/978-94-015-1252-7_2

Covington W (1975) Altitudinal variation of chlorophyll concentration and reflectance of the bark of *Populus tremuloides*. *Ecology* 56: 715–20. doi.org/10.2307/1935507

Filella I, Peñuelas J (1999) Altitudinal differences in UV absorbance, UV reflectance and related morphological traits of Quercus ilex and Rhododendron ferrugineum in the Mediterranean region. *Plant Ecology* 145(1): 157–65. doi.org/10.1023/A:1009826803540

Holman J, Hughes J, Fensham R (2011) Origins of a morphological cline between *Eucalyptus melanophloia* and *Eucalyptus whitei*. *Australian Journal of Botany* 59: 244–52. doi.org/10.1071/BT10209

Jurasinski G, Kreyling J (2007) Upward shift of alpine plants increases floristic similarity of mountain summits. *Journal of Vegetation Science* 18: 711–18. doi.org/10.1658/1100-9233(2007)18[711:USOAPI]2.0.CO;2

O'Sullivan OS, Weerasinghe K, Evans J, Egerton J, Tjoelker M, Atkin O (2013) High-resolution temperature responses of leaf respiration in snow gum (*Eucalyptus pauciflora*) reveal high-temperature limits to respiratory function. *Plant, Cell and Environment* 36: 1268–84. doi.org/10.1111/pce.12057

Pescador D, De Bello F, Valladares F, Escudero A (2015) Plant trait variation along an altitudinal gradient in Mediterranean high mountain grasslands: Controlling the species turnover effect. *PLoS ONE* 10: e0118876. doi.org/10.1371/journal.pone.0118876

Royer D, McElwain J, Adams J, Wilf P (2008) Sensitivity of leaf size and shape to climate within *Acer rubrum* and *Quercus kelloggii*. *New Phytologist* 179: 808–17. doi.org/10.1111/j.1469-8137.2008.02496.x

Speed J, Austrheim G, Hester A, Mysterud A (2012) Elevational advance of alpine plant communities is buffered by herbivory. *Journal of Vegetation Science* 23: 617–25. doi.org/10.1111/j.1654-1103.2012.01391.x

Turkis S, Ozbucak T (2010) Foliar resorption and chlorophyll content in leaves of *Cistus creticus* L. (Cistaceae) along an elevational gradient in Turkey. *Acta Botanica Croatica* 69: 275–90.

Wang Y, Čufar K, Eckstein D, Liang E (2012) Variation of maximum tree height and annual shoot growth of smith fir at various elevations in the Sygera Mountains, southeastern Tibetan Plateau. *PLoS ONE* 7: e31725. doi.org/10.1371/journal.pone.0031725

Zhong M, Wang J, Liu K, Wu R, Liu Y, Wei Z, Pan D, Shao X (2014) Leaf morphology shift of three dominant species along altitudinal gradient in an alpine meadow of the Qinghai-Tibetan Plateau. *Polish Journal of Ecology* 62: 639–48. doi.org/10.3161/104.062.0409

Plastic responses to environmental stressors: Biosynthesis of anthocyanins increases in *Eucalyptus pauciflora* and *Richea continentis* with elevation

Gregory Gauthier-Coles, Cynthia Turnbull, Ray Zhang, Tanja Cobden

Abstract

As a subclass of the flavonoid family of secondary metabolites, anthocyanins have been studied closely in recent years for their multitudinous protective properties in plants. Described as 'nature's Swiss army knife', anthocyanins have putative antioxidant, anti-inflammatory, anti-microbial, anti-cancer, photo-protective and colligative characteristics, which have made them a subject of much scientific interest. Anthocyanins are the primarily expressed group of flavonoids in angiosperms such as *Eucalyptus pauciflora* and *Richea continentis*. These species of plants are widely found in the Australian alpine environment of the Kosciuszko National Park, a region that is particularly susceptible to climate change. In studying anthocyanin concentration in the leaves of *E. pauciflora* and *R. continentis* in parts of the Kosciuszko National Park, anthocyanin concentration was found to positively correlate with elevation. In accordance with previous scientific research, anthocyanin accumulation greatly increased in leaves showing evidence of pathogen attack. In spite of methodological limitations, results of this study support the notion of plasticity in the expression of enzymes involved in anthocyanin biosynthesis. Furthermore, these findings may have implications in climate change modelling in relation to plant species distribution of the Australian alpine region and in conservation ecology.

Introduction

The Australian alpine and subalpine environment is characterised by several species of plants, most notably *Eucalyptus pauciflora*, otherwise known as the snow gum (Figure 1A). This robust species of the Australian flora has a large altitudinal range (Boland and McDonald 2006)—they can be found at sea level and make up the alpine tree line in the Australian Alps (William and Potts 1996). *E. pauciflora*'s ability to grow and even thrive in relatively inhospitable alpine and subalpine environments is not only due to its cold hardiness but its tolerance for drought and other stressors (Close et al. 2010). Indeed, the same can be said for many plant species in these environments, such as the much less studied *Richea continentis* (Figure 1B).

The properties that allow such species of plants to exist at high elevation can be explained by important morphological and physiological differences. For example, the seedlings of *E. pauciflora* have been observed to predominantly grow and regenerate on the southern side of the tree. This protects the seedlings from high levels of sunlight and frost damage (Ball et al. 1991). *E. pauciflora* also has a significant capacity for acclimation in its stomatal and photosynthetic responses to changes in temperature and illumination (Körner and Cochrane 1985). The production of biomolecules that have the necessary protective qualities for the survival of plants in alpine and subalpine regions is another example of a set of adaptations that are found in plants like *E. pauciflora* and *R. continentis*. One such group of biomolecules is the flavonoid family of secondary metabolites.

Flavonoids comprise more than 9,000 metabolites, and were an essential biochemical innovation in plants that, well over 500 million years ago, had left the marine environment and needed to adapt to, and endure, severe terrestrial stressors such as higher UV irradiation and greater temperature extremes (Williams and Grayer 2004; Mouradov and Spangenberg 2014). These biomolecules share a common metabolic pathway. The first step in flavonoid synthesis involves the production of aromatic amino acids, such as phenylalanine, using the shikimate pathway. The resulting products of this pathway are then deaminated, hydroxylated and decarboxylated, and a reaction involving three molecules of malonyl-coenzyme A results in the biosynthesis of the first flavonoid, producing either a chalcone or a stilbene (Grotewold 2006). The next step for the majority of flavonoids involves the formation of a C ring, and many subsequent reactions lead to the large diversity of these biomolecules (Hernández and Van Breusegem 2010).

Figure 1: Plant species examined in this study growing in the Kosciuszko National Park: A) *Eucalyptus pauciflora*, B) *Richea continentis*.

Source: Authors' photographs.

A feature of flavonoids is that, unlike proteins, they do not contain nitrogen. Furthermore, due to the many reaction pathways necessary, their synthesis consumes significant amounts of energy. This can be useful under situations, for example, cold temperatures and high light, in which the light reactions of photosynthesis produce a lot of ATP and NADPH; however, the Calvin cycle operates slowly, leading to a buildup of this ATP and NADPH. The flavonoid biosynthesis pathway requires some of the same precursors as the Calvin cycle and uses large amounts of ATP and NADPH. Therefore, it has been hypothesised that flavonoids act as 'energy escape valves' when the Calvin cycle operates slowly (Hernández and Van Breusegem 2010).

This mechanism is considered useful in instances where the photosynthetic machinery is operational to a disproportionate degree to that of the Calvin cycle reactions. The bifurcation found in the glycolysis pathway allows for glyceraldehyde-3-phosphate and dihydroxyacetone phosphate metabolites to be diverted to the shikimate pathway; the photosynthesis reactions can therefore be uncoupled from the Calvin cycle. This prevents the accumulation of the photosynthetic products ATP and NADPH, and in a way that does not exploit nitrogen reserves, which can be strained in certain conditions (Shirley 1996): the upregulation of enzymes involved in flavonoid biosynthesis requires nitrogen in the form of amino acids. This phenomenon may occur in situations where the leaves of a plant are exposed to certain abiotic stressors, such as high illumination and cold temperatures.

One of the most important subclasses of metabolites belonging to the flavonoid family in angiosperms is that of the anthocyanins. Comprising some 500 different structures, anthocyanins are multifunctional chemicals in plants characterised by a host of qualities (Guo et al. 2008). One of these is the ability to pigment the leaves and flowers of plants. Anthocyanins also reflect blue-green and ultraviolet light and therefore act as sunscreens in plants, specifically in palisade and spongy mesophyll cells (Tattini et al. 2005). This is very important in allowing plants to avoid photo-oxidative stress, specifically to photosynthetic machinery when it is exposed to high solar radiation. Stratmann et al. (2007) have partially elucidated the mechanisms by which ultraviolet radiation activates signals that recruit the transcription apparatus for certain structural genes (i.e. those coding for enzymes involved in anthocyanin biosynthesis). This appears to be an example of physiological plasticity in plants. The sunscreen effect, however, is not the only photo-protective feature of anthocyanins, as they are also potent antioxidants (Rozema et al. 1997).

Reactive oxygen species (ROS) are the products of many biological impacts, such as photo-oxidation. Hydrogen peroxide and superoxide radicals are ROS that are readily created as a result of exposure to high solar radiation, and can react together to form hydroxyl radicals and anions as part of the Haber-Weiss reaction (Haber and Weiss 1932). These products cause serious damage to cellular components including DNA. This same reaction can be catalysed by transition metal ions, namely ferric ions—this is known as the Fenton reaction (Fenton 1894). Since ferric ions naturally occur in various parts of plant cells (including the nucleus), serving a variety of physiological roles, the production of ROS via the Fenton reaction is a serious problem. Flavonoids, including anthocyanins located in the nucleus, exert an anti-oxidative effect indirectly by chelating transition metals, thereby preventing their involvement in Fenton reactions (Melidou et al. 2005). Although this may not be an example of direct anti-oxidative action, vacuolar flavonoids (mostly anthocyanins and proanthocyanidins) have true antioxidant capacities (Hernandez et al. 2009). When the vacuole that contains these flavonoids is breached due to some form of mechanical injury (such as feeding by insect or animal herbivores), anthocyanins and proanthocyanidins are released and neutralise ROS by donating their electrons (Gould et al. 2002).

There is also some evidence that anthocyanins promote cold hardiness in plants that accumulate them; however, this remains an area of contention (Chalker-Scott 1999). Although not fully understood, Christie et al. (1994) found that the anthocyanin production pathway involves cold regulation, or *cor*, genes. It therefore seems that anthocyanin biosynthesis may be induced by cold temperature, but it is not understood if this has an explicit function (i.e. conferring frost resistance in plant tissues). One hypothesis states that anthocyanins, acting as solutes, have an osmotic effect that lowers the freezing point of water in vacuoles (Chalker-Scott 1999). Another hypothesis, not necessarily mutually exclusive with the former, suggests that anthocyanin production is induced by cold temperatures due to the fact that ROS are longer lived in such circumstances (Kramer et al. 1991). This proposal is further supported by evidence that anthocyanin biosynthesis is prevented in cold conditions where visible or UV-B light is absent (Janda et al. 1996; Oren-Shamir and Levi-Nissim 1997).

Anthocyanins are involved in osmotic stress conditions, not only occurring in plants subjected to cold temperatures but also drought. Many plants that are capable of withstanding drought express high levels of anthocyanins. Bahler et al. (1991) found that cultivars of pepper that expressed far greater quantities of anthocyanins (and were purple in colour as a result) had a greater tolerance for drought conditions. *Craterostigma wilmsii* and *Xerophyta viscosa* were also observed to have far higher concentrations of anthocyanins in their roots under drought conditions as opposed to when the plants were fully hydrated.

In summarising the well-established and speculated functions and 'cross-resistances' of anthocyanins, the term 'Swiss army knife' assigned by Kevin S. Gould seems appropriate for such a group of biomolecules (Gould et al. 2002). Furthermore, many of the properties of anthocyanins clearly show a capacity for plasticity. From the biosynthetic pathways that are photo-inductive to those that involve *cor* genes, expression of gene encoding anthocyanin biosynthesis enzymes is determined by numerous genetic and environmental factors. Following on from previous research reported in the literature, the aim of this study was to investigate changes in the anthocyanin concentration in the leaves of *E. pauciflora* and *R. continentis* at different elevation. Given the evidence for plasticity in anthocyanin biosynthesis in response to colder temperatures and increased UV light intensities, it was hypothesised that the anthocyanin concentrations in both these species would increase with elevation in the Australian Alps.

Materials and methods

Sampling of *E. pauciflora* and *R. continentis*

Five samples of *E. pauciflora* and of *R. continentis* were collected at different elevations (range: 1,600 to 2,000 m) in the Charlotte Pass and Mount Stillwell area. Each sample consisted of three replicates (i.e. three branches were selected from the same plant sample). Another *E. pauciflora* sample was taken to analyse anthocyanin expression in leaves that showed evidence of previous damage from herbivory and/or pathogen(s) (Figure 2). Samples were picked, placed in bags and labelled with time of day, elevation and temperature. Elevation was estimated based on information from topographic maps, and temperature determined using an alcohol thermometer. Samples were not selected based on aspect (exposure to sunlight); however, leaves damaged due to herbivory or the presence of obvious pathogens were excluded. Two samples of *R. continentis* were selected at the same elevation; however, one was found growing adjacent to a snow patch whereas the other was found growing without surrounding snow.

Preparation for thin layer chromatography

One leaf of each *E. pauciflora* and two leaves of each *R. continentis* were weighed and then ground with approximately 0.5 g of glass powder in a mortar and pestle. Because anthocyanin content was clearly different in leaves of different developmental stages (Figure 2), especially for *E. pauciflora*, we always collected the third youngest leaf from each branch. For the leaf showing signs of damage from herbivores and/or pathogens (Figure 2C), which was of a mature stage, we used a mature *E. pauciflora* leaf as a control. The pulverised leaves were then deposited in Eppendorf tubes and an ethanol–water mixture (70:30 v/v) was added. Each solution was then centrifuged for 2.5 minutes. The amount of solvent added related to the mass of the leaves used in each replicate and is shown in the following equation (where V is the volume of solvent and m is the mass of leaf extract):

$$V = m \times 2$$

Figure 2: Leaves of *Eucalyptus pauciflora* and *Richea continentis* analysed for anthocyanin content: A) *E. pauciflora*, B) *R. continentis*, C) *E. pauciflora* leaf with obvious damage and anthocyanin accumulation from herbivore and/or pathogen damage.

Note: Magnification bar represents 2 cm in panels A and B and 1.2 cm in C.

Source: Authors' photographs.

Thin layer chromatography

The supernatant of each tube was spotted onto silica-coated thin layer chromatography (TLC) plates (Merck, Darmstadt, Germany) in a way that all three replicates of each sample were allocated to a given plate. A consistent volume of solution was added to each spot; a content equivalent to two 5 cm lengths of a glass capillary tube, 1 mm diameter, was deposited. These spots were placed along a line that was 1 cm above the edge of the plate. After around 5 minutes of drying time, plates were placed in a running buffer solution prepared with ethyl acetate, distilled water, acetone and formic acid (18:18:3:1) for approximately 10 minutes. Plates were then removed and allowed to dry for a further 10 minutes.

Determination of anthocyanin density

One photograph was taken for all the *E. pauciflora* plates and one for all the *R. continentis* plates, under the same illumination, using an iPhone 6 camera. The images were then analysed using ImageJ software. For each photograph, colours were inverted to allow the anthocyanin band to be demarcated. The anthocyanin band was clearly visible as a pink band at the bottom of each plate. A standardised surface area was then used for each photograph and the bands of anthocyanin highlighted. The selected bands were then analysed and an integrated density value obtained for each band. A background integrated density value was also determined (based on a selection of an area of a TLC plate that was blank and unstained) and a net integrated value was determined by subtracting the background integrated density from the anthocyanin band integrated density.

Data analysis

The net integrated densities were averaged for each sample and standard deviations were obtained. Linear regressions tests were then used for the *E. pauciflora* data set and the *R. continentis* data set. A t-test was used for the *E. pauciflora* pathogenic vs non-pathogenic data set, and to compare the *R. continentis* samples from the same elevation differing by proximity to snow. All data analysis was conducted using Microsoft Excel.

Results

Separation of plant pigments from *Eucalyptus pauciflora* and *Richea continentis* leaves using thin layer chromatography showed a clear separation of anthocyanins from other pigments, including chlorophylls and carotenoids. Quantification of the relative band intensities using ImageJ showed that there was a significant positive correlation between elevation and anthocyanin content in both *E. pauciflorus* (Figure 3A) and *R. continentis* (Figure 3B).

An additional observation made during the collection of leaves was that some leaves that showed signs of attack by herbivores and/or pathogens showed intense red pigmentation (Figure 2C). We therefore compared the relative anthocyanin content of a diseased and a healthy leaf, both of a mature developmental stage. As shown in Figure 4, there was a significant increase in anthocyanin content in the diseased leaf.

When collecting samples of *R. continentis*, we observed strong red pigmentation in leaves of specimens growing close to a patch of snow, as opposed to specimens growing in areas without snow at the same elevation (2,000 m). A quantitative comparison of relative anthocyanin content of leaves from each plant showed a higher anthocyanin content in the leaves of the plant growing close to the snow patch compared to the plant growing in a snow-free area. However, this difference was not statistically significant ($P>0.05$; Figure 5).

A)   B)

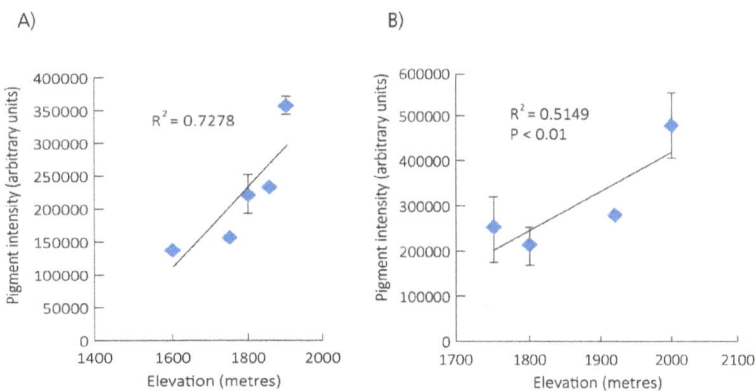

Figure 3: Relationship between elevation and relative anthocyanin concentration in leaves of A) *Eucalyptus pauciflora* and B) *Richea continentis*, analysed by linear regression.

Note: Data points indicate means and standard errors, n=3.

Source: Authors' data.

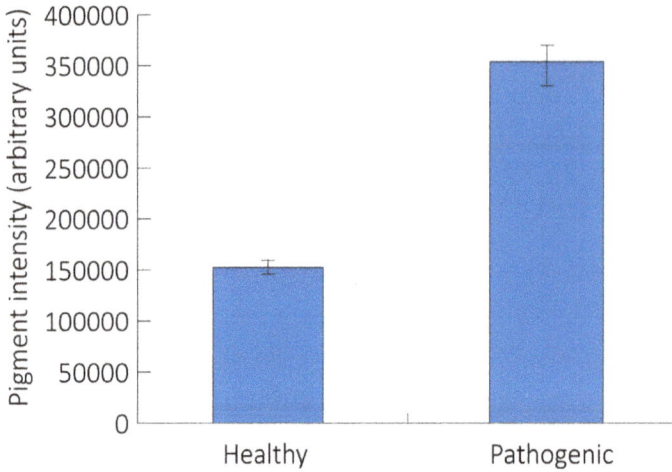

Figure 4: Relative anthocyanin concentrations of healthy vs pathogen-infected *Eucalyptus pauciflora* leaves, *P*<0.001 (tailed, two sample homoscedastic t-test). Data show means and standard errors of three individual leaves.

Source: Authors' data.

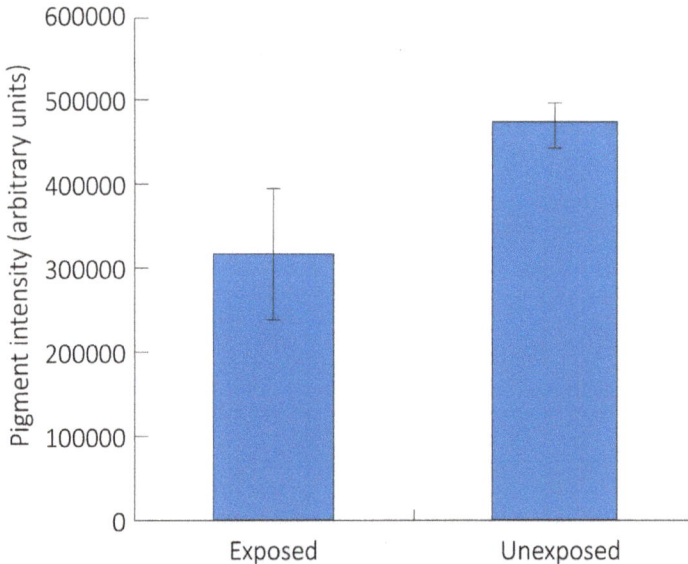

Figure 5: Relative anthocyanin concentrations of *Richea continentis* leaves growing in an exposed area without snow or in an unexposed area close to a snow patch, *P*>0.05 (tailed, two sample homoscedastic t-test). Data show means and standard errors of three individual leaves.

Source: Authors' data.

Discussion

In investigating anthocyanin expression in the leaves of *E. pauciflora* and *R. continentis* at different elevation, the hypothesis that a positive correlation exists between these two factors was supported. The prediction that anthocyanin concentration would be greater in a pathogen-infected *E. pauciflora* compared to a healthy leaf was also supported. The prediction that *R. continentis* leaves growing in an unexposed area compared to those growing in an exposed area at the same elevation could not be supported. These results are largely in keeping with the literature, which finds that plants do produce higher concentrations of anthocyanins at lower temperatures (and, therefore, as elevation increases) (Kramer et al. 1991).

Despite the fact that the R^2 values of the linear regressions performed for each plant against elevation were not very high, the low *P*-values suggest that a relationship between these two variables exists and that further studies should be carried out. Apart from increasing the sample sizes for each experiment (anthocyanin production vs elevation; anthocyanin concentration in pathogen vs healthy *E. pauciflora* leaves; and anthocyanin concentration in exposed vs unexposed *R. continentis* leaves), this experiment could have been improved in several ways. For instance, we did not control for the level of exposure to sunlight in plants. A thermometer was used, but the data was discarded as only one measurement was taken for each sample site at different times of the day. The use of elevation as the independent variable, and as a substitution for temperature changes, may not have been reliable since it did not control for discrepancies in microclimates. For example, the Charlotte Pass area lies in a valley that acts as a cold sink. Temperatures in this valley are frequently below those at higher elevation, which may have affected the way the sampled plants expressed anthocyanins in their leaves. Although the anthocyanin band on the TLC plates was fairly easily identifiable, no standards were used and it is possible that other phytochemicals with similar polarity aggregated at the same level as the anthocyanins. This problem, combined with the use of the ImageJ program, was not the ideal method of determining anthocyanin content in the leaves.

The methodology employed could have therefore been improved in several ways. Sampling could have been selected over a large vertical range and in different areas of the Kosciuszko National Park. This would help control for microclimatic changes at different elevations. Temperature data could have been used if multiple recordings were taken at each sample site and over several days (preferably over a year-long period). Likewise, the use of an illuminometer may have helped control for sun exposure; however, readings would have needed to be taken at different times of day and over several weeks. Anthocyanin standards could have been used in the TLC assays but an even better analytical technique would comprise the use of high pressure liquid chromatography coupled with mass spectrometry. This method would allow a more precise concentration to be determined, and would elucidate on the ratio of individual anthocyanin constituents.

Although this study had some serious methodological limitations, some important implications can be drawn. The data showed significant plasticity in how these plants express anthocyanins in response to altitudinal gradients. Plasticity was also observed in plant leaves that were infected, whereby anthocyanin concentration is vastly increased in the presence of pathogens. There was also some evidence (albeit statistically insignificant) that suggested that *R. continentis* produces higher levels of anthocyanins close to snow, which may have enhanced light exposure due to reflectance from the snow, and this should be investigated further. The plastic nature of anthocyanin expression in these plant species has implications in climate change research, especially modelling involved in plant distributions in the Australian Alps. These findings and, more broadly, the area of research concerning the production of anthocyanins and other flavonoids may be of interest to conservationists involved in restoration efforts. For instance, plants that are capable of expressing high levels of anthocyanins in response to a variety of stressors (i.e. drought, frost, physical damage) may be good candidates for environmental rehabilitation projects.

Future research should primarily consist of further investigating the aim of this study but in a way that controls for the aforementioned variables and with much greater sample sizes. A study with a high level of ecological validity would share a similar design to the experiment outlined in this report, but would be conducted over a year to investigate seasonal differences in anthocyanin concentration of leaves. Alternatively, many variables could be more easily controlled for in a laboratory setting. A multifactorial laboratory experiment could assess anthocyanin concentrations in *E. pauciflora* and *R. continentis* by individually manipulating temperature and luminescence. Although the idea of anthocyanins acting as super-coolant molecules is a controversial one, and despite the fact that this was not observed in the 'exposed vs unexposed' *R. continentis*, a future experiment could test for frost avoidance by using anthocyanin structural gene knockouts and wild-type plants at sub-zero temperatures and search for frost damage. These gene knockout experiments could also target the *cor* genes that are suspected to be involved in the expression of enzymes that participate in anthocyanin biosynthesis. Field trials could also be designed to test how well plants that produce high levels of anthocyanins (like *E. pauciflora* and *R. continentis*) help rehabilitate areas damaged by or susceptible to climate change and other destructive human activities.

Acknowledgements

We would like to thank Professor Ulrike Mathesius from the Research School of Biology at The Australian National University for her invaluable contributions to this study.

References

Bahler B, Steffen K, Orzolek M (1991) Morphological and biochemical comparison of a purple-leafed and green-leafed pepper cultivar. *HortScience* 26: 736.

Ball M, Hodges V, Laughlin G (1991) Cold-induced photoinhibition limits regeneration of snow gum at tree-line. *Functional Ecology* 5: 663–8. doi.org/10.2307/2389486

Boland D, McDonald M (2006) *Forest trees of Australia*. CSIRO Publishing, Collingwood.

Chalker-Scott L (1999) Environmental significance of anthocyanins in plant stress responses. *Photochemistry and Photobiology* 70: 1–9. doi. org/10.1111/j.1751-1097.1999.tb01944.x

Christie P, Alfenito M, Walbot V (1994) Impact of low-temperature stress on general phenylpropanoid and anthocyanin pathways: Enhancement of transcript abundance and anthocyanin pigmentation in maize seedlings. *Planta* 194: 541–9. doi.org/10.1007/BF00714468

Close D, Davidson N, Churchill K, Corkrey R (2010) Establishment of native *Eucalyptus pauciflora* and exotic *Eucalyptus nitens* on former grazing land. *New Forests* 40: 143–52. doi.org/10.1007/s11056-010-9189-9

Fenton H (1894) LXXIII. Oxidation of tartaric acid in presence of iron. *Journal of the Chemical Society, Transactions* 65: 899–910. doi.org/10.1039/CT8946500899

Gould K, McKelvie J, Markham K (2002) Do anthocyanins function as antioxidants in leaves? Imaging of H_2O_2 in red and green leaves after mechanical injury. *Plant, Cell and Environment* 25: 1261–9. doi.org/10.1046/j.1365-3040.2002.00905.x

Grotewold E (2006) *The science of flavonoids*. Springer Publishing, New York. doi.org/10.1007/978-0-387-28822-2

Guo J, Han W, Wang M (2008) Ultraviolet and environmental stresses involved in the induction and regulation of anthocyanin biosynthesis: A review. *African Journal of Biotechnology* 7: 4966–72.

Haber F, Weiss J (1932) Uber die Katalyse des Hydroperoxydes. *Die Naturwissenschaften* 20: 948–50. doi.org/10.1007/BF01504715

Hernández I, Alegre L, Van Breusegem F, Munné-Bosch S (2009) How relevant are flavonoids as antioxidants in plants? *Trends in Plant Science* 14: 125–32. doi.org/10.1016/j.tplants.2008.12.003

Hernández I, Van Breusegem F (2010) Opinion on the possible role of flavonoids as energy escape valves: Novel tools for nature's Swiss army knife? *Plant Science* 179: 297–301. doi.org/10.1016/j.plantsci.2010.06.001

Janda T, Szalai G, Páldi E (1996) Chlorophyll fluorescence and anthocyanin content in chilled maize plants after return to a non-chilling temperature under various irradiances. *Biologia Plantarum* 38: 625–7. doi.org/10.1007/BF02890623

Körner C, Cochrane P (1985) Stomatal responses and water relations of *Eucalyptus pauciflora* in summer along an elevational gradient. *Oecologia* 66: 443–55. doi.org/10.1007/BF00378313

Kramer G, Norman H, Krizek D, Mirecki R (1991) Influence of UV-B radiation on polyamines, lipid peroxidation and membrane lipids in cucumber. *Phytochemistry* 30: 2101–8. doi.org/10.1016/0031-9422(91)83595-C

Melidou M, Rignakos K, Galaris D (2005) Protection against nuclear DNA damage offered by flavonoids in cells exposed to hydrogen peroxide: The role of iron chelation. *Free Radical Biology and Medicine* 39: 1591–600. doi.org/10.1016/j.freeradbiomed.2005.08.009

Mouradov A, Spangenberg G (2014) Flavonoids: A metabolic network mediating plants adaptation to their real estate. *Frontiers in Plant Science* 5: 620. doi.org/10.3389/fpls.2014.00620

Oren-Shamir M, Levi-Nissim A (1997) UV-light effect on the leaf pigmentation of *Cotinus coggygria* 'Royal Purple'. *Scientia Horticulturae* 71: 59–66. doi.org/10.1016/S0304-4238(97)00073-3

Rozema J, van de Staaij J, Björn L, Caldwell M (1997) UV-B as an environmental factor in plant life: Stress and regulation. *Trends in Ecology and Evolution* 12: 22–8. doi.org/10.1016/S0169-5347(96)10062-8

Shirley B (1996) Flavonoid biosynthesis: 'New' functions for an 'old' pathway. *Trends in Plant Science* 1: 377–82.

Stratmann, J, Stelmach B, Weiler E, Ryan C (2007) UVB/UVA radiation activates a 48 kDa myelin basic protein kinase and potentiates wound signaling in tomato leaves. *Photochemistry and Photobiology* 71: 116–23. doi.org/10.1562/0031-8655(2000)0710116SIPUUR2.0.CO2

Tattini M, Guidi L, Morassi-Bonzi L, Pinelli P, Remorini D, Degl'Innocenti E, Giordano C, Massai R, Agati G (2005) On the role of flavonoids in the integrated mechanisms of response of *Ligustrum vulgare* and *Phillyrea latifolia* to high solar radiation. *New Phytologist* 167: 457–70. doi.org/10.1111/j.1469-8137.2005.01442.x

Williams CA, Grayer R (2004) Anthocyanins and other flavonoids. *ChemInform* 35: 539–73. doi.org/10.1002/chin.200447250

Williams K, Potts B (1996) The natural distribution of *Eucalyptus* species in Tasmania. *Tasforests* 8: 39–165.

Circadian rhythms and leaf characteristics of subalpine *Eucalyptus pauciflora*: The effects of environmental exposure on photosynthetic machinery

Christine Mauger, Julia Hammer, Chen Liang, Cameron McArthur, Angela Stoddard

Abstract

This study explored whether snow gums (*Eucalyptus pauciflora*) in a subalpine region use circadian rhythms for photosynthesis and whether age and exposure affect photosynthetic machinery. Fv/Fm, SPAD and specific leaf area (SLA) were measured in individual leaves to assess variation in leaf traits. Fv/Fm was not significantly different between exposed and sheltered trees, but old leaves had significantly higher Fv/Fm values than young leaves. SPAD was higher in exposed trees and old leaves had significantly higher chlorophyll content. SLA was significantly lower in old leaves, and sheltered leaves had significantly higher SLA than exposed leaves. Increase in SLA with exposure was greater in young leaves. Branches from sheltered and exposed *E. pauciflora* were harvested then placed in light and dark treatments. Stomatal conductance was tested at various times to determine if leaf stomata were opening without access to sunlight. At the time of this experiment, there had been no previous known studies on circadian rhythms in *E. pauciflora*. Conductance rates were higher in the light treatment, and across this treatment it was higher at 11 am than at night. There was no effect of exposure on conductance rates in the light treatment. In the dark treatment, exposed leaves had lower conductance rates than sheltered leaves. For sheltered and exposed leaves, conductance was greater at dawn than at 11 pm and greater again at 11 am than at dawn, suggesting *E. pauciflora* do use circadian rhythms for photosynthesis.

Introduction

Photosynthesis is the conversion of light energy to chemical energy, stored as sugars or other compounds in plants. For photosynthesis to take place, the plant undergoes gas exchange, which is controlled by the stomata within the leaf's epidermis. In subalpine to alpine areas such as the Australian Snowy Mountains low temperatures occur during the spring snow melt. There is a short growing season with water and light availability restricted. A plant's ability to undergo photosynthesis is dependent on resource availability, and so in these areas plants need to adapt to deal with such stresses. Most organisms do not respond specifically to sunrise, but do await the dawn. When external time cues are absent, many of these diurnal rhythms continue, indicating there is an endogenous biological circadian clock (McClung 2006). Circadian rhythms demonstrate an approximate 24-hour period or cycle. They are generated internally within the organism, and continue under free-running conditions like constant light or darkness, and temperature (de Dios et al. 2009). The circadian clock influences gas exchange in a plant by preparing for dawn and dusk cycles (de Dios et al. 2009) and is thought to regulate a large number of plant processes. Circadian rhythms assist plant growth, meaning they can give plants a competitive advantage. This has been demonstrated in the model plant *Arabidopsis thaliana*. Dodd et al. (2005) found that in wild type and mutants of *A. thaliana*, plants that had a clock period matching the environment contained more chlorophyll. They were also able to fix more carbon, as well as grow faster and survive better, showing that circadian control gives plants an advantage. The diurnal variation in photosynthetic gas exchange is of interest to botanists, ecologists, conservationists and climate modellers, who seek to understand the effect future changes in climate may have on ecosystems (de Dios et al. 2009). Stomatal conductance is a measurement of the rate that carbon dioxide enters or water vapour exits the stomata. Because stomata are open during photosynthesis, stomatal conductance can be used as a proxy to measure photosynthetic activity. Typically, stomatal conductance increases with light up until around midday. In the middle of the day, the demand for water from the atmosphere increases, and to avoid unnecessary water loss, stomatal conductance declines (de Dios et al. 2009).

Leaf morphology can also affect photosynthetic characteristics. Leaf characteristics may vary depending on exposure, orientation and age in ways that alter carbon gain patterns. Specific leaf area (SLA) is the ratio of leaf area to dry mass, measured in cm²/g. Typically, thicker leaves have a lower SLA and have more sclerenchyma. Sclerenchyma is a plant tissue that provides strength, making a plant hardy. Thinner leaves have higher SLA with more mesophyll cells, which are specialised for photosynthesis (Steane et al. 2015). SPAD is a measurement of chlorophyll content within leaves and is thus linked to photosynthetic ability. The amount of sunlight that a leaf absorbs is a function of the chlorophyll; therefore, chlorophyll content affects photosynthetic potential (Gitelson et al. 2003; Curran et al. 1990; Filella et al. 1995). Fv/Fm is a measurement of chlorophyll fluorescence and an indicator of photosystem health. Chlorophyll fluorescence is commonly used to measure plant stress and is linked to photosynthetic efficiency (Patankar et al. 2013).

E. pauciflora is the dominant species found at the tree line in the Australian Alps (Brookhouse et al. 2008) and is a keystone species occurring in the Kosciuszko National Park. While there have been some studies done on circadian rhythms in Eucalypt species (de Dios et al. 2013), there has been none on *E. pauciflora*. Given future climate change predictions for the subalpine and alpine regions, it is important to understand how this species' photosynthetic machinery works in varying conditions.

The aim of this study was to see if stomatal conductance varied over a 24-hour period, and if it was dependant on the leaf's position, its exposure and age in subalpine *E. pauciflora* without any environmental cues. Our main hypothesis with the above factors considered was that exposed snow gums would show less circadian rhythm regulation, in terms of stomatal conductance, than sheltered snow gums.

We assessed variation in leaf level gas exchange over a 24-hour period in *E. pauciflora*. SLA, chlorophyll content (SPAD) and the efficiency of photosynthetic machinery (Fv/Fm) were assessed. In addition to time of day, we also considered the amount of variation within and between plants, looking at the age of the leaves and their position on the plant. When looking at variation in leaf traits, we hypothesised Fv/Fm would be lower in exposed plants and that young leaves would have higher Fv/Fm. We hypothesised that age and exposure would affect chlorophyll content (SPAD) with it being lower in exposed leaves, and lower in older leaves. Our last hypothesis was that SLA would be lower in exposed leaves.

Methods

Study site

Figure 1: Site map showing sampling sites. Sheltered branches and leaves were harvested on the Rainbow Lake path in subalpine woodland and are marked with X (n=5). Exposed trees that were sampled along the road between Charlotte Pass and Rainbow Lake are marked with red pins (n=7).

Source: Created by the authors in GPS visualiser (www.gpsvisualizer.com).

Sheltered branches were collected close to the Rainbow Lake walking trail (−36.371114°S, 148.479931°E) in Kosciuszko National Park, New South Wales, Australia. Selected exposed trees were single standing so they were fully exposed to environmental conditions (Figure 1).

Circadian rhythm measurements

One branch from each sheltered tree (n=5) and exposed tree (n=7) was collected to be used for the light and dark treatments. The branches had healthy leaves and were of a size that could be broken to create two branches. Branches were numbered (1 to 5 and 6 to 12) and stems were cut far down. Branches were individually wrapped in wet paper towel and placed in a plastic bag containing water. Samples were taken to the lab to prepare for stomatal conductance measurements in light and dark treatments. A bucket was filled halfway with cold tap water. Each branch was placed in the water, where a cut was made below water level to reduce the risk of embolisms, caused by allowing air to enter the vascular tissue. While still below the water, the branch was placed into a zip-lock bag and labelled with tree number, and then into a tall cup so it could stand up straight. If the bottom of the branch came out of water during movement, it was cut again below the water level. Once all branches were prepared for treatment and placed in cups, they were randomly allocated into two groups (for light and dark treatment) with one branch from each tree put into each group. Light treatment branches were put under constant artificial lighting (JB HPS400 N4208 grow lamp). Dark treatment branches were placed in a constantly dark, windowless room so light was unable to enter.

Each branch had a single leaf that was marked with tape and tree number. Branches were left in treatment conditions for at least eight hours before conductance measurements were taken. Stomatal conductance and temperature measurements were taken using the Decagon SC-1 Leaf Porometer for marked leaves at 11 pm and then 5 am and 11 am the following morning. Measurements of dark treatment branches were conducted using a green light setting on a head torch, to avoid affecting results.

Leaf trait measurements

Leaf samples were taken back to the lab for Fv/Fm, SPAD and SLA measurements. Four individual leaves were taken from the same trees where branches were removed, with a combination of young, old, north-facing and south-facing leaves (old = older than last years' growth, young = last years' growth). Selected sheltered trees were those close together, with the aim of minimising differences in exposure levels. Leaves were numbered and placed into a zip-lock bag. Fv/Fm was measured using a Plant Efficiency Analyser. Dark adaptation was conducted for 20 minutes on each individual leaf, using leaf clips before receiving a light flash to centre of leaf and measuring Fv/Fm. Chlorophyll content was measured with a Konica Minolta Chlorophyll Meter (SPAD-502) in SPAD units as an average of three measurements for each leaf. Leaves were photographed and had leaf area measured using image analysis with ImageJ. Leaves were dried overnight and each individual leaf had SLA (cm^2/g) measured using an electronic balance. SLA is calculated by dividing the area of the leaf in cm^2 by the dry leaf mass in grams.

Statistical analyses

The data were tested using analysis of variance (ANOVA) to look at the effects age and exposure had on SLA, Fv/Fm, SPAD and also on stomatal conductance.

Results

The ANOVA (Table 1) yielded the following results: old leaves have significantly higher ($P<0.001$) Fv/Fm than young leaves but both are quite healthy. Fv/Fm was slightly higher in old exposed leaves compared to old sheltered leaves (Figure 2a). Photosynthetic machinery efficiency (Fv/Fm) was not significantly different ($P=0.673$) between exposed and sheltered trees, or between age and exposure ($P=0.535$). Old leaves have significantly higher ($P<0.001$) chlorophyll content (SPAD). When comparing sheltered and exposed sites, SPAD was higher, but not significantly ($P=0.200$), in leaves from exposed trees (Figure 2b). Old leaves have significantly lower ($P<0.001$) SLA than young leaves, and sheltered leaves have significantly higher SLA than exposed leaves ($P<0.001$). The increase in SLA with exposure is also significantly ($P=0.010$) greater in young leaves (Figure 2c).

Table 1: ANOVA summary table.

Source	SLA	Fv/Fm	SPAD (chlorophyll)
Age	***	***	***
exposure	***	ns	ns
Age*exposure	*	ns	ns

Note: * $P<0.05$, ** $P<0.01$, *** $P<0.001$, ns = not significant.
Source: Authors' data.

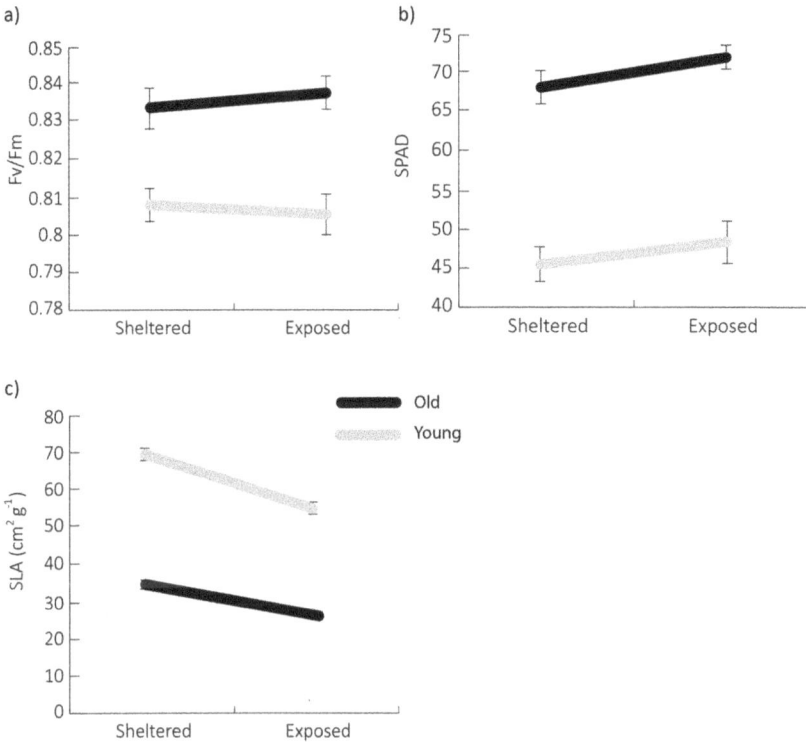

Figure 2: a) Comparison of average Fv/Fm in young and old leaves within sheltered and exposed sites, b) Comparison of average SPAD in young and old leaves within sheltered and exposed sites, c) Comparison of average SLA in young and old leaves within sheltered and exposed sites.

Note: Error bars are standard error.

Source: Authors' data.

Stomatal conductance of sheltered leaves (Figure 3a) increased overnight under the dark treatment, nearly doubling from a mean of 53 mmol m^{-2}s^{-1} at 11 pm to a mean of 96 mmol m^{-2}s^{-1} at 11 am. Stomatal conductance rates of exposed leaves (Figure 3b) increased overnight under dark treatment with the mean conductance starting at 40 mmol m^{-2}s^{-1} at 11 pm and 74 mmol m^{-2}s^{-1} at 11 am.

ANOVA results also show that plants in the dark treatments had significantly (P=0.014) lower conductance rates in exposed leaves than in sheltered leaves. For both exposures in the dark, ANOVA results show conductance was greater at dawn than at 11 pm, and significantly (P=0.024) greater again at 11 am compared to dawn.

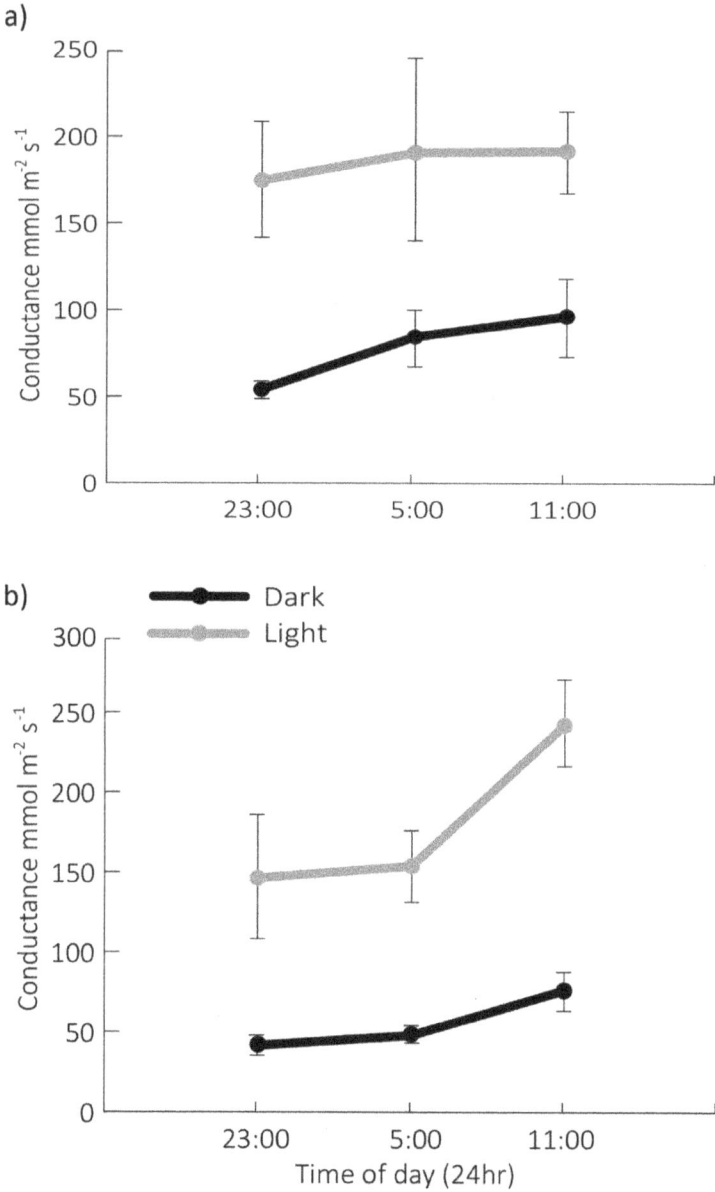

Figure 3: a) Mean conductance measurements of sheltered leaves under continuous dark and light treatments. Conductance was measured at 23:00, 05:00 and 11:00 hours, b) Mean conductance measurements of exposed leaves under continuous dark and light treatments. Conductance was measured at 23:00, 05:00 and 11:00 hours.

Note: Error bars are standard error.

Source: Authors' data.

Under the light treatment stomatal conductance increased from 11 pm to 5 am (by 16 mmol m^{-2}s^{-1}), but not from 5 am to 11 am for sheltered leaves. The conductance rate was much higher for sheltered leaves in the light treatment (mean 174–190 mmol m^{-2}s^{-1}); however, dark sheltered leaves showed a greater increase in stomatal conductance over time.

When looking at light and exposure effects, results from the ANOVA showed that conductance was significantly ($P<0.001$) higher in the light, and across light treatments it was significantly higher at 11 am than at night ($P=0.035$). Under the light treatment, stomatal conductance of exposed leaves increased overnight from a mean of 145 mmol m^{-2}s^{-1} at 11 pm to a mean of 240 mmol m^{-2}s^{-1} at 11 am.

Light treatment leaves from exposed trees appeared to capitalise more on light around midday than those from sheltered sites; however, when considering just those plants in the light treatment, there is no significant effect of exposure ($P=0.642$), or period ($P=0.156$) on conductance.

Discussion

Leaf traits

We hypothesised Fv/Fm would be lower in exposed plants, with the need of investing energy into sclerenchyma over mesophyll cells, and that young leaves would have higher Fv/Fm. The ANOVA (Table 1) results showed old leaves have significantly higher ($P<0.001$) Fv/Fm than young ones but both are quite healthy. A leaf with Fv/Fm above 0.8 is considered a healthy leaf. This was not expected as typically younger leaves are better at photosynthesising, but it may be due to our selection methods where old was considered to be anything older than the previous year's growth. The ANOVA shows that photosynthetic machinery efficiency (Fv/Fm) was not significantly different ($P=0.673$) between exposed and sheltered trees, or between age and exposure ($P=0.535$), going against our hypothesis.

We hypothesised that age and exposure would affect chlorophyll content (SPAD) with it being lower in exposed leaves, and lower in older leaves. The ANOVA (Table 1) showed that old leaves in fact have significantly higher ($P<0.001$) chlorophyll content (SPAD), going against our hypothesis. This is likely to be that energy is going into plant structural growth. When comparing sheltered and exposed sites, SPAD was higher but not significantly (P=0.200) in leaves from exposed trees (Figure 2b), also going against our hypothesis. SPAD may be lower in sheltered trees as less light is able to reach the leaves as they are in close proximity to each other so chlorophyll content is lower.

The ANOVA (Table 1) showed that old leaves have significantly lower (P<0.001) SLA than young leaves and that sheltered leaves have significantly higher SLA than exposed leaves (P<0.001). The increase in SLA with exposure is also significantly (P=0.010) greater in young leaves. These results support our hypothesis that SLA is lower in exposed leaves. High SLA is typical of thinner or less dense, younger leaves. An explanation for lower SLA in exposed trees is that more extreme environmental conditions may have led to the development of more hardy structural tissues for leaf protection.

Circadian rhythm

Our main hypothesis was that exposed snow gums would show less circadian rhythm regulation, in terms of stomatal conductance, than sheltered snow gums. Leaves from the dark treatment (both exposed and sheltered trees) became more active, with stomata conductance increasing from 11 pm to 11 am, indicating there is circadian rhythm in *E. pauciflora*. ANOVA results show that in plants in the dark treatments there was significantly (P=0.014) lower conductance in exposed leaves than in sheltered leaves. These results support our hypothesis that exposed snow gums would show less circadian rhythm, in terms of stomatal conductance, than sheltered snow gums. Light treatment leaves from exposed trees appeared to capitalise more on light around midday than those from sheltered sites. However, when considering just those plants in the light treatment, there is no significant effect of exposure (P=0.642), or period (P=0.156) on conductance; when looking at circadian rhythm, it is the dark treatment that needs to be considered.

Many previous studies looking at circadian rhythms in plants show clear diurnal patterns in photosynthesis and respiration. Conductance peaking at midday and lowering around midnight in the dark treatment indicates the use of circadian rhythm (Patankar et al. 2013). Stomatal conductance rates were higher in leaves from sheltered trees at the measurement times of 11 pm and 5 am for both treatments compared to leaves from exposed trees (Figures 3a and 3b). The 11 pm measurement was also higher in the dark treatment for sheltered leaves than exposed. Interestingly, the light treatment leaves from exposed trees appeared to capitalise more on light around midday than those from sheltered sites with a difference in 50 mmol m^{-2}s^{-1}.

For a circadian rhythm to have adaptive benefit, there must be consistency between average days. Circadian rhythms persist although external cues are absent. The large increase in stomatal conductance for exposed trees in the light treatment could be due to midday being the time there is the most solar light available for energy production and they are adapted to utilising the light, being fully exposed to it.

Generally, photosynthetic gas exchange is strongly driven by the environment. However, many studies have now provided evidence that circadian regulation plays a large, important role in many plant species (de Dios et al. 2009). This could explain why the conductance rate in exposed trees in the light treatment was so high. This could also explain why the conductance rates for the sheltered and exposed leaves in the light treatment only had a slight increase between 11 pm and 5 am. Being constantly in front of a light treatment, they had reached the optimal conductance rate, and the circadian clock at that time did not allow for more stomata to open in the risk of losing water.

In subalpine regions, there are many stresses to plants and so they may have limited opportunity to photosynthesise each day. Opening stomata before light hits the plant gives them a competitive advantage against others, limiting water loss that may occur from evaporation during the day.

Further studies

Sheltered leaves and branches that were harvested came from an area (Rainbow Lake) that had been burnt in the 2003 fires. Exposed leaves came from an unburnt area. It would be interesting to test if some of these results were due to rehabilitation. When looking at this factor, it would be important to look both at exposed trees from a similar area and at sheltered trees from an unburnt area for comparison. Testing leaves from the east and west sides of trees may also give different results.

If this study was to be replicated, it would be ideal to have the same sample number for exposed and sheltered sites as the numbers were uneven (n=7 and n=5) in this experiment, which could possibly skew the results. It is important to note that there were a number of people working on this project with a handover, possibly leading to variations in procedures and measurements causing possible human error.

As this was a study with a small sample size, at one time of the year, it would be ideal to test if the same trees (sheltered and exposed) went through any seasonal changes in relation to stomatal conductance and if the quality of photosynthetic machinery varied. If these factors affect growth, survival and competitive advantage, a longitudinal study would provide data either proving or disproving this, and it may vary during seasons.

Seasonal patterns of plant physiology recorded at midday in the Northern Hemisphere show that photosynthesis is often lowest right after snow melt in spring. It reaches its peak around the middle of the growing season in mid-summer, and decreases again towards autumn (Defoliart et al. 1988; Starr et al. 2000, 2008). Around autumn is when leaves start hardening and the health of the leaves deteriorate (Oberbauer and Starr 2002). This is interesting as the results of this study show some of the older leaves tested had greater photosynthetic efficiency and greater chlorophyll content. It would be interesting to see if this is a general seasonal pattern that also occurs in the Snowy Mountains for other evergreen Eucalypt species, as this was a one-time experiment conducted at the start of summer. Whilst undergoing statistical analysis for SLA, there was one outlier removed, which was of young age and exposed. Removing this outlier did not affect the pattern of results.

Conclusions

As *E. pauciflora* is a keystone species of the Kosciuszko National Park, understanding the diurnal physiology of snow gums is important and can help when considering conservation projects, including producing models. It would be particularly helpful for the use of predicting and understanding the effects of climate change on the Australian Snowy Mountain flora. Climate change may lead to a longer growing season, with warmer summers. Change in temperatures may induce a shift in the composition of plant communities in the region.

In relation to global climate change predictions in the Snowy Mountains of Australia, an increase in ambient temperature will likely result in greater water loss through stomatal conductance. This will affect plants, ability to photosynthesise and limit their growth. Plants that use circadian rhythm to control stomatal conductance, opening stomata during hours of darkness, will have a competitive advantage over those that do not. This allows those using circadian rhythms to capitalise on morning light without losing too much water. However, without endogenous control to close the stomata and stop photosynthesising during the hottest time of the day, plants may lose large amounts of water and fall victim to heat stress.

Acknowledgements

Thanks to our project supervisor Adrienne Nicotra, and to Sarah Stock, Hannah Zurcher, Ming-Dao Chia and Tess Walsh-Rossi for a smooth project handover and for supplying data for sheltered leaves/branches.

References

Brookhouse M, Lindesay J, Brack C (2008) The potential of tree rings in *Eucalyptus pauciflora* for climatological and hydrological reconstruction. *Geographical Research* 46: 421–34. doi.org/10.1111/j.1745-5871.2008.00535.x

Curran PJ, Dungan JL, Gholz HL (1990) Exploring the relationship between reflectance red edge and chlorophyll content in slash pine. *Tree Physiology* 7: 33–48. doi.org/10.1093/treephys/7.1-2-3-4.33

de Dios VR, Diaz-Sierra R, Goulden ML, Barton CVM, Boer MM, Gessler A, Ferrio JP, Pfautsch S, Tissue DT (2013) Woody clockworks: Circadian regulation of night-time water use in *Eucalyptus globulus*. *New Phytologist* 200: 743–52. doi.org/10.1111/nph.12382

de Dios VR, Hartwell J, Hall A (2009) Ecological implications of plants' ability to tell the time. *Ecology Letters* 12: 583–92. doi.org/10.1111/j.1461-0248.2009.01295.x

Defoliart L, Griffith M, Chapin FS, Jonasson S (1988) Seasonal patterns of photosynthesis and nutrient storage in *Eriophorum vaginatum* L., an arctic sedge. *Functional Ecology* 2: 185–94. doi.org/10.2307/2389694

Dodd AN, Salathia N, Hall A, Kévei E, Tóth R, Nagy F, Hibberd JM, Millar AJ, Webb AA (2005) Plant circadian clocks increase photosynthesis, growth, survival, and competitive advantage. *Science* 309(5734): 630–3. doi.org/10.1126/science.1115581

Filella I, Serrano I, Serra J, Peñuelas J (1995) Evaluating wheat nitrogen status with canopy reflectance indices and discriminant analysis. *Crop Sci* 35: 1400–5. doi.org/10.2135/cropsci1995.0011183X003500050023x

Gitleson AA, Gritz Y, Merzlyak MN (2003) Relationships between leaf chlorophyll content and spectral reflectance and algorithms for non-destructive chlorophyll assessment in higher plant leaves. *Journal of Plant Physiology* 160: 271–82. doi.org/10.1078/0176-1617-00887

McClung CR 2006. Plant circadian rhythms. *Plant Cell* 18: 792–803. doi.org/10.1105/tpc.106.040980

Oberbauer SF, Starr G (2002) The role of anthocyanins for photosynthesis of Alaskan arctic evergreens during snowmelt. *Advances in Botanical Research* 37: 129–45. doi.org/10.1016/S0065-2296(02)37047-2

Patankar R, Mortazavi B, Oberbauer SF, Starr G (2013) Diurnal patterns of gas-exchange and metabolic pools in tundra plants during three phases of the arctic growing season. *Ecology and Evolution* 3: 375–88. doi.org/10.1002/ece3.467

Starr G, Oberbauer SF, Ahlquist LE (2008) The photosynthetic response of Alaskan tundra plants to increased season length and soil warming. *Arctic Antarctic, and Alpine Research* 40: 181–91. doi.org/10.1657/1523-0430(06-015)[STARR]2.0.CO;2

Starr G, Oberbauer SF, Pop E (2000) Effects of lengthened growing season and soil warming on the phenology and physiology of *Polygonum bistorta*. *Global Change Biology* 6: 357–69. doi.org/10.1046/j.1365-2486.2000.00316.x

Steane DA, Potts BM, McLean E, Collins L, Prober MS, Stock WD, Vaillancourt RE, Byrne M (2015) Genome-wide scans reveal cryptic population structure in a dry-adapted eucalypt. *Tree Genetics & Genomes* 11: 2500–13. doi.org/10.1007/s11295-015-0864-z

Thermal acclimation of photosynthesis and respiration in *Eucalyptus pauciflora* of varying growth temperatures in Kosciuszko National Park

Caitlin McLeod, Cynthia Turnbull, Gregory Gauthier-Coles, You Zhou, Ainsley Maurer, Ray Zhang, Tanja Cobden, Yvonne Yong

Abstract

The processes of photosynthesis and respiration in plants are largely responsible for levels of atmospheric CO_2 globally. Predicting future output and uptake of carbon by plants is therefore crucial for developing accurate climate change models. The effect of temperature acclimation on photosynthesis and respiration rates in plants is currently not considered in global carbon models. Therefore, current terrestrial carbon models may be overestimating carbon release from plants as acclimation to higher growth temperatures in many species leads to a fall in respiration and an increase in the temperature at which photosynthesis is optimised. In this study, the ability of photosynthesis and respiration in snow gums (*Eucalyptus pauciflora*) to acclimate was tested. Leaf samples were taken from trees at four different elevations, and rates of photosynthesis and respiration were measured at 25°C using a licor 6400 Gas Exchange System. Temperatures of each collection site (growth temperatures) were measured hourly over three days using temperature and humidity data loggers (ibuttons). Results were analysed by comparing growth temperature of the samples to the rate of respiration and photosynthesis at 25°C. Lower temperatures correlated to low rates of photosynthesis and higher rates of respiration, when measured at the common temperature of 25°C. These results indicate that respiration and photosynthesis in snow gums does acclimate to ambient temperature, and this allows for the development of more accurate climate change models.

Introduction

Respiration is the cellular process of the breakdown of energy stores to produce carbon for structural purposes, useable energy (ATP) and reducing agents (NADH) (Atkin and Tjoelker 2003). This process is necessary for the growth and maintenance of tissues and produces CO_2 as a by-product (Amthor 2000). Photosynthesis is the production of sugars from CO_2 and water, which can then be used in respiration. Both of these processes are undergone due to the functioning of many enzymatic processes, and these are temperature sensitive (Cooper 2000). Therefore, the efficiency of respiration and photosynthesis is dependent on how far the ambient air temperature is from the optimal functioning temperature of the leaf.

Atmospheric carbon dioxide levels have increased by nearly a third since the industrial revolution and are continuing to increase, and are responsible for much of the present and future planet warming. Currently, plants release about 60 gigatonnes of carbon into the atmosphere annually through respiration, which equates to approximately 50 per cent of all CO_2 released globally (King et al. 2006; Atkin and Tjoelker 2003). Plants also fix about two-thirds more carbon during photosynthesis than is released during respiration (Atkin et al. 2000).

The rates at which plants photosynthesise and respire therefore heavily affects the accuracy of climate change models. Understanding plant responses to increases in temperature that are inevitable due to climate change is necessary for predicting rates of photosynthesis and respiration in plants in the future. Having a better understanding of this effect will improve the accuracy of current climate change models.

There is evidence that there is acclimation (the process of changing to survive in environments of different temperatures (Stillman 2003)) of both photosynthesis and respiration to increases in temperature in some plant functional types, including woody plants (Liang et al. 2013). Forested area, of which woody plants are the dominant plant functional type, covered approximately 35 per cent of the earth's land surface area in 1997 (current figures are unknown) and thus contribute a very large proportion to the net carbon release and uptake due to plant respiration and photosynthesis (Hansen et al. 2010; Curtis and Wang 1998). Therefore, due to their ability to acclimate, it can be predicted that there would be no change in release or fixation of CO_2 in woody plants as a result of a future warmer climate.

Snow gums (*Eucalyptus pauciflora*) are a dominant plant species in the Australian Alps and exist over a large area of south-eastern Australia. These trees exist over a range of elevations and temperatures but originate from the same gene pool (Slayter and Morrow 1977). Therefore, if snow gum leaf samples from different locations exhibit different responses when placed under identical conditions, they have changed in response to their environment, and thus acclimation has occurred.

The aim of this study was to determine if acclimation of respiration and photosynthesis occurs in the snow gums of the Australian Alps. It was hypothesised that samples from warmer environments would undergo respiration at a slower rate than samples from colder environments, and that respiration would increase with decreasing growth temperature when tested at 25°C. It was also hypothesised that samples from warmer environments would photosynthesise at a greater rate than colder environments and that rate of photosynthesis would decrease with decreasing growth temperature when tested at 25°C.

Materials and method

Collection

Samples were collected at four different sites in the Kosciuszko National Park: Lake Jindabyne, elevation 900 m ± 30 m; Kosciuszko information centre, elevation 1,220 m ± 30 m; 50 m along walking track to Rainbow Lake, elevation 1,620 m ± 30 m; Charlotte Pass, elevation 1,830 m ± 30 m. Samples were selected from three healthy, mature, north-facing snow gums. A north-facing branch with at least one fully mature and damage-free (disease, insect or otherwise) leaf was cut and placed in a bucket of water. The end was then recut underwater to prevent embolisms. The time that the branches were cut was recorded. A temperature and humidity ibutton was attached to each tree within 5 cm of the sample cut. Samples were transported back to the lab immediately and placed near a window in the sun to allow photosynthesis to continue.

Analysis

In the lab, the largest healthy leaf from each branch was labelled and placed in a Licor 6400 Gas Exchange System and tested under photosynthetic conditions (conditions: Block T: 25°C; CO_2R: 400 ppm; PAR: 1,500 μmol

$m^{-2}s^{-1}$; humidity: around 60 per cent). Rates of photosynthesis (amount of CO_2 uptake/area of leaf/second) were recorded when readings were steady. Foil was then placed over the leaf for a minimum of 30 minutes. Leaves were then measured again using the Licor under respiration conditions (Block T: 25°C; CO_2R: 400 ppm; PAR: 0; humidity: around 60 per cent). As with photosynthesis, rates of respiration were recorded when readings were steady.

Dry leaf mass (used to convert rates of respiration and photosynthesis from an area basis to a mass basis to account for leaf thickness) were measured by cutting a 2 cm × 3 cm rectangle (size of the Licor cuvette) out of the centre of the leaf. The rectangle was placed in an envelope and this was put between two sheets of cardboard and microwaved (microwave brand: Tiffany) three times on medium/low power for 2 minutes. The sample was weighed and placed in a bag with silica gel to prevent hydration. The weight was used to convert photosynthetic and respiration readings from the Licor from $\mu mol\ CO_2\ m^{-2}s^{-1}$ to $nmol\ CO_2\ g^{-1}s^{-1}$.

The ibuttons were removed three days after being attached to the sampled tree. Temperature and humidity recordings were taken once per hour over 48 hours. The average of these readings were calculated and used as the 'growth temperature' for each site. These averages were considered representations of the actual growth temperatures of the samples and they exhibited the relationship between elevation and temperature according to Hopkin's Bioclimatic law (which is used to predict how much colder higher elevations will be than lower elevations at a given latitude).

Results

Both the day and night air temperature decreased with elevation over the 48-hour period of sampling for the four sample sites (Figure 1). The average of temperatures taken hourly over a 48-hour period in the environment where the sample was taken was used as the growth temperature for each elevation. Growth temperatures were Jindabyne 28.59°C, Kosciuszko Information Centre 16.48°C, Rainbow Lake walking track 14.88°C, Charlotte Pass 14.12°C.

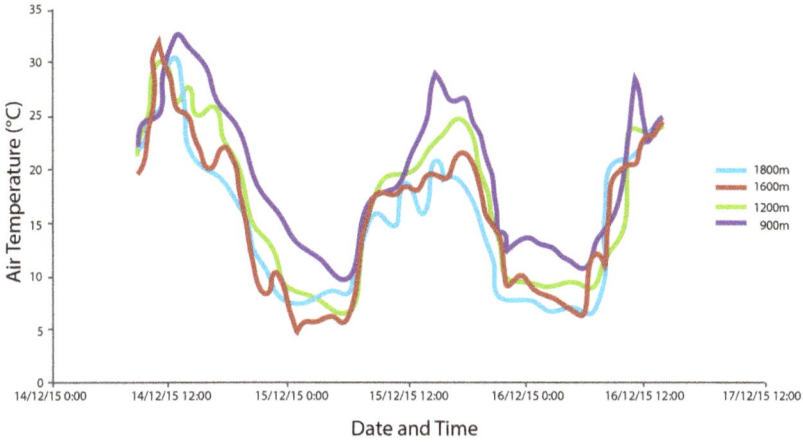

Figure 1: Air temperature over a 48-hour period for each of the sample collection sites, measured using temperature data loggers, recording once per hour.

Source: Authors' data.

When tested at 25°C, the rate of respiration in *E. pauciflora* with growth temperature: 28.59°C was 3.39 ± 0.645 µmol m^{-2}s^{-1}; 16.48°C, 2.85 ± 0.422 µmol m^{-2}s^{-1}; 14.88°C, 5.67 ± 0.330 µmol m^{-2}s^{-1}; and 14.12°C, 5.02 ± 2.88 µmol m^{-2}s^{-1}. Overall, there was an increase in the rate of respiration as growth temperature increased (Figure 2a).

When tested at 25°C, the rate of photosynthesis in *E. pauciflora* with growth temperature: 28.59°C was 20.9 ± 2.69 µmol m^{-2}s^{-1}; 16.48°C, 21.4 ± 1.88 µmol m^{-2}s^{-1}; 14.88°C, 16.4 ± 2.25 µmol m^{-2}s^{-1}; and 14.12°C, 15.0 ± 2.72 µmol m^{-2}s^{-1}. Overall, there was a decrease in rate of photosynthesis as growth temperature increased (Figure 2a).

The increase in photosynthesis and decline in respiration with growth temperature when measured at the common 25°C, leads to a decline in respiration as a fraction of photosynthesis with increasing temperature (Figure 2c). Hence, with increasing growth temperature, *E. pauciflora* is able to retain more carbon per unit of leaf area.

There was no discernible trend between leaf mass per area (LMA) and temperature, with all sites having similar LMA, except Rainbow Lake at 14.22°C, which had a lower LMA.

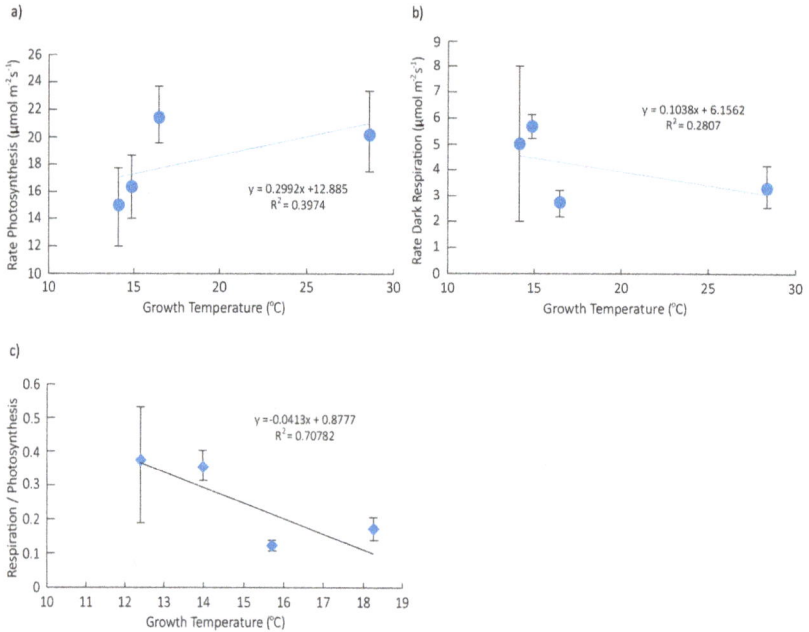

Figure 2: a) Rate of photosynthesis in *Eucalyptus pauciflora* measured at 25°C for the four growth temperatures, b) Rate of respiration in *Eucalyptus pauciflora* measured at 25°C for the four growth temperatures, c) Respiration as a fraction of photosynthesis (respiration/photosynthesis), measured at the common temperature of 25°C, for each of the growth temperatures.

Note: Error bars show the standard deviation in the 3 samples from each site.

Source: Author's data.

Discussion

As expected, temperature decreased with elevation. This allowed comparisons to be made between respiration and photosynthesis for snow gums growing at a range of temperatures. The trend in the photosynthesis results suggests that plants grown in warmer environments photosynthesise at a greater rate at 25°C than plants grown in colder environments, and that the rate of photosynthesis at 25°C decreases with decreasing growth temperature. This agrees with the proposed hypothesis. The opposite relationship was found for respiration, with plants grown in warmer environments respiring at a slower rate at 25°C than plants grown in colder environments, and that respiration rate at 25°C increases with decreasing growth temperature. This also agrees with the proposed hypothesis. Taken together, the increased photosynthesis and reduced respiration of plants growing at higher temperatures results in the decline of the respiration to photosynthesis ratio as growth temperature increases. This has implications on carbon models as it suggests that as plants acclimate to higher temperatures, the amount of carbon lost by the plant to the atmosphere will be less than that expected if acclimation did not occur.

Of interest, the variability in respiration at the lowest growth temperature was much more than for any of the other growth temperatures. This may be due to the lowest growth temperature corresponding to the highest elevation. One argument is that a more harsh and less homogenous environment on the top of the mountain may have led to more variable respiration rates between individual trees, as other environmental variables apart from temperature may be influencing respiration.

As 25°C is much warmer than the average temperatures in the Australian Alps, the conditions under which the samples were tested were most similar to the growth conditions of the warmest sample site. As photosynthesis usually reaches a peak at a temperature close to the growing temperature (Yamori et al. 2013), it is not surprising that photosynthesis was higher in trees growing at temperatures closer to the measured temperature. This means the photosynthetic apparatus has altered to function best in the temperature at which the tree exists. This indicates that photosynthetic acclimation occurs in snow gums.

A similar conclusion can be drawn from the respiration results. As leaves from different sample sites react differently under identical conditions, they must have altered in response to their environment to attain optimal function. Thus, it is likely that snow gums grown at different temperatures respire at a near identical rate when measured at those growth temperatures. This indicates that respiratory acclimation, like photosynthesis, also occurred in snow gums.

Some differences in the growth conditions of the samples could not be controlled and so should be noted when analysing the results of this experiment. The site with growth temperature 16.48°C (Information Centre) was a very forested, sheltered area. The trees grew very tall (likely due to the protection from extremes that the surrounding large trees and relatively flat surrounds provided) and thus we were only able to reach leaves on small trees that lacked large, fully mature leaves. At the site with growth temperature 14.88°C (Rainbow Lake) every tree was burned from bushfires in 2003 so it was difficult to find large, mature leaves. In both of these cases, we chose the most mature-looking leaves we could find.

The trees at Charlotte Pass (14.12°C) were all in flower or about to flower. This was not the case in trees in other locations. Therefore, these trees were investing energy into making flowers and so it is likely that the rate that was attained for dark respiration was not actually the rate of dark respiration due to investing energy into reproduction.

An indication of the divergence in environmental conditions, other than temperature, that could affect physiological processes between the four sites is evident in the LMA results. LMA usually increases in cold and hot environments (Poorter et al. 2009). An increase in LMA with increasing elevation was expected, as the cold and harsh environment (snow covered in winter) of the mountain top would induce thicker, more robust leaves to withstand those conditions. However, there was no clear trend between LMA and temperature/elevation. Therefore, local environmental conditions, rather than temperature gradient associated with the elevation gradient, were impacting on leaf morphology and presumably also impacting on photosynthesis and respiration rates.

There has been some evidence in the literature to suggest that acclimation of plant respiration to temperature does not occur (Dewar et al. 1999; Dillaway and Kruger 2011; Liang et al. 2013). Because of this, it was predicted that the increased temperatures, inevitable due to climate change, would result in an increase in the rate of plant respiration and thus an increase in CO_2 release, and this would then speed up climate change further, which would result in more CO_2 release and the cycle would go on. This idea was incorporated into the climate change models.

Most evidence now is suggesting thermal acclimation of respiration and photosynthesis does occur (Atkin et al. 2000; Bunce 2007; King et al. 2006) and this study supports that evidence. It suggests that future atmospheric CO_2 levels may not be as high as previously thought and this idea is now being incorporated into current climate change models. The accuracy of these models is improving, and this allows society to rely more heavily on the models for future planning.

Recent evidence is suggesting that other environmental factors have a much greater effect on rates of photosynthesis and respiration in plants than temperature, including leaf nitrogen content, water availability and levels of carbohydrates in the plants (Dillaway and Kruger 2011; Lewis et al. 2011). Further research into the effects of these factors as well as how predicted environmental changes will affect these factors would also improve the predictions of the levels of CO_2 and temperature in the future. More accurate knowledge on this would allow for more accurate climate change models.

What controls respiration and photosynthesis physiologically is still widely unknown (Gonzalez-Meler et al. 2004) and, ultimately, having a complete understanding of how these processes work would give a greater insight into how plants are likely to respond to the pressures of the future climate.

Acknowledgements

The authors thank Owen Atkin, Andrew Scafaro and Nur Abdul Bahar for their guidance and assistance.

References

Amthor J (2000) The McCree-de Wit-Penning de Vries-Thornley respiration paradigms: 30 years later. *Annals of Botany* 86: 1–20. doi. org/10.1006/anbo.2000.1175

Atkin I, Holly C, Ball M (2000) Acclimation of snow gum (*Eucalyptus pauciflora*) leaf respiration to seasonal and diurnal variation in temperature: The importance of changes in the capacity and temperature sensitivity of respiration. *Plant, Cell & Environment* 23: 15–26. doi.org/10.1046/j.1365-3040.2000.00511.x

Atkin O, Tjoelker M (2003) Thermal acclimation and the dynamic response of plant respiration to temperature. *Trends in Plant Science* 8: 343–51. doi.org/10.1016/S1360-1385(03)00136-5

Bunce J (2007) Direct and acclimatory responses of dark respiration and translocation to temperature. *Annals of Botany* 11: 67–73. doi. org/10.1093/aob/mcm071

Cooper GM (2000) Metabolic energy. In: *The Cell: A Molecular Approach,* 2nd edn. ASM Press, Washington, DC; Sinauer Associates, Sunderland, MA. doi.org/10.1007/s004420050381

Curtis P, Wang X (1998) A meta-analysis of elevated CO_2 effects on woody plant mass, form, and physiology. *Oecologia* 113: 299–313.

Dewar R, Medlyn B, Mcmurtrie R (1999) Acclimation of the respiration/photosynthesis ratio to temperature: Insights from a model. *Global Change Biology* 5: 615–22. doi.org/10.1046/j.1365-2486.1999.00253.x

Dillaway D, Kruger E (2011) Leaf respiratory acclimation to climate: Comparisons among boreal and temperate tree species along a latitudinal transect. *Tree Physiology* 31: 1114–27. doi.org/10.1093/treephys/tpr097

Gonzalez-Meler M, Taneva L, Trueman R (2004) Plant respiration and elevated atmospheric CO_2 concentration: Cellular responses and global significance. *Annals of Botany* 94: 647–56. doi.org/10.1093/aob/mch189

Hansen M, Stehman S, Potapov P (2010) Quantification of global gross forest cover loss. *Proceedings of the National Academy of Sciences of the United States of America* 107: 8650–5. doi.org/10.1073/pnas.0912668107

King A, Gunderson C, Post W, Weston D, Wulischleger S (2006) Plant respiration in a warmer world. *Science* 312: 536–7. doi.org/10.1126/science.1114166

Lewis J, Phillips N, Logan B, Hricko C, Tissue D (2011) Leaf photosynthesis, respiration and stomatal conductance in six *Eucalyptus* species native to mesic and xeric environments growing in a common garden. *Tree Physiology* 31: 997–1006. doi.org/10.1093/treephys/tpr087

Liang J, Xia J, Liu L, Wan S (2013) Global patterns of the response of leaf-level photosynthesis and respiration in terrestrial plants to experimental warming. *Journal of Plant Ecology* 6: 437–47. doi.org/10.1093/jpe/rtt003

Ow L, Whitehead D, Walcroft A, Turnbull M (2008) Thermal acclimation of respiration but not photosynthesis in *Pinus radiata*. *Functional Plant Biology* 35: 448–61. doi.org/10.1071/FP08104

Poorter H, Niinemets U, Poorter L, Wright I, Villar R (2009) Causes and consequences of variation in leaf mass per area (LMA): A meta-analysis. *New Phytologist* 182: 565–88. doi.org/10.1111/j.1469-8137.2009.02830.x

Slayter R, Morrow P (1977) Altitudinal variation in the photosynthetic characteristics of snow gum, *Eucalyptus pauciflora* Sieb. Ex Spreng. I. Seasonal changes under field conditions in the Snowy Mountains area of south-eastern Australia. *Australian Journal of Botany* 25: 1–20. doi.org/10.1071/BT9770001

Stillman J (2003) Acclimation capacity underlies susceptibility to climate change. *Science* 301: 65. doi.org/10.1126/science.1083073

Yamori W, Hikosaka K, Way D (2013) Temperature response of photosynthesis in C3, C4, and CAM plants: Temperature acclimation and temperature adaptation. *Photosynthesis Research* 119: 101–17. doi.org/10.1007/s11120-013-9874-6

Why do alpine plants grow where they do?

This project led by Dr Susanna Venn was not written up as a full report but is included here as an abstract.

Abstract

In an alpine landscape, sharp gradients in climate factors such as wind, snow accumulation and temperature, and physical factors such as topography, aspect and soil, all interact to create environmental gradients, along which various groups of plants will find suitable for habitat. These environmental factors act as filters to exclude species with unviable physiological limitations from entering and persisting in different communities. It is these environmental filters and sharp gradients in habitat suitability that produce the distinct patterns in plant communities that we see in the alpine landscape. However, species *per se* do not respond to environmental factors; rather, it is how they function within their community and interact with other species that determines where they grow and how successful they are. And thus, an understanding of plant functional traits becomes important for understanding how species and communities will respond to differences and changes in these environmental factors/filters.

Plant functional traits, or morphological traits such as height, size, growth form, specific leaf area and leaf chlorophyll content, are the physical characteristics of species that reveal how plants capture and conserve resources, and how they interact with their environment and each other. As such, plant functional traits underpin the functioning of ecosystems (McGill et al. 2006). For example, plant height can indicate a species' overall competitive ability (tall plants can overtop neighbouring plants as they compete for light and space) as well as being an indirect measurement for biomass, lateral spread, rooting depth and leaf size. Specific leaf area (SLA), the ratio of the area of a fresh leaf to its dry mass, can indicate how plants allocate resources;

low SLA values correspond to relatively long-lived leaves with high investments in defences or other structural adaptations to cope with harsh conditions (Leishman and Westoby 1992). Ecologists can use trait-based approaches to predict where certain types of species might be found across landscapes or environmental gradients, or even how various plant communities might respond to climate change via changes to the local driving environmental factors (Venn et al. 2011).

In this project, we hiked to and surveyed several contrasting plant communities near Charlotte Pass. We compared them floristically and functionally by measuring various plant functional traits. In this way, we were able to determine which ecological processes (as inferred from their traits) might be driving each of the communities we sampled, and gave us an idea of how these communities might respond to future changes in environmental drivers.

References

Leishman M, Westoby M (1992) Classifying plants into groups on the basis of associations of individual traits: Evidence from Australian semi-arid woodlands. *Journal of Ecology* 80(3): 417–24. doi. org/10.2307/2260687

McGill BJ, Enquist, BJ, Weiher E, Westoby M (2006) Rebuilding community ecology from functional traits. *Trends in Ecology and Evolution* 21: 178–85. doi.org/10.1016/j.tree.2006.02.002

Venn SE, Green K, Pickering, CM, Morgan, JW (2011) Using plant functional traits to explain community composition across a strong environmental filter in Australian alpine snowpatches. *Plant Ecology* 212: 1491. doi.org/10.1007/s11258-011-9923-1

Mortality or more-tile-ity: Tiling response to temperature and humidity extremes in *Agrotis infusa*

Hannah Zurcher, Chen Liang, Ming-Dao Chia, Sarah Stock, Tess Walsh-Rossi

Abstract

The bogong moth (*Agrotis infusa*) is one of Australia's most famous insects, due in part to its spectacular tiling behaviour. The moths have great social and historical significance to Indigenous Australians from large portions of south-eastern Australia, and have been adopted as an icon by many modern groups and artists. These moths aestivate in rocky environments over summer, but often interact with human structures, bringing them spectacularly to the attention of the general public. Despite this, the exact mechanisms driving bogong moth tiling are unknown. The migratory lifestyle of the bogong moth enables moths to avoid the worst of temperature and humidity extremes during the Australian summer. However, conditions in the alpine and subalpine caves in which the moths prefer to aestivate are still intensely warm and dry, but with nights much cooler than the lower altitudes that the moths have left. While the purpose of the moths' tiling behaviour is currently unknown, given these conditions it has been proposed that moths may tile to conserve moisture, heat or both. We exposed groups of moths to temperature extremes, while measuring tiling percentage and effectiveness, and also exposed moths either singly or in groups to extreme humidity, while measuring the same. Contrary to predictions, moths demonstrated both higher temperatures and higher moisture loss when tiling.

Introduction

The bogong moth (*Agrotis infusa*) exhibits tiling behaviour during aestivation. The precise drivers for this behaviour are currently unknown, but could possibly be related to temperature, time of year, relative humidity, diet, elevation, migration triggers, social responses, diurnal cues or predation responses (Common 1954).

Moths aestivate in caves or crevices in boulder fields, generally in alpine or subalpine locations. As aestivation is an energy-intensive process (moths generally have between 55 per cent and 60 per cent body fat before entering aestivation (Common 1954)), moths will need to conserve resources wherever possible. With this in mind, the tiling behaviour may possibly be an adaptation intended to conserve moisture by creating a more humid environment, to minimise water loss via gas exchange.

As the preferred moth aestivation sites are in alpine environments, night and day temperatures vary immensely, especially compared to the more stable environments in which they overwinter. The tiling behaviour may possibly be an adaptation that regulates temperatures during alpine thermal extremes. The unreliability of water access to aestivating moths, and the potential for parasitism by nematodes from available water sources (Welch 1963), means that water conservation is also an important factor.

With warming trends persisting globally, the temperature and humidity profile of the Australian Alps is in danger of dramatic change. If the tiling behaviour of bogong moths functions to stabilise temperature or humidity in aestivating groups, conditions may change rapidly enough that the tiling behaviour becomes useless or, worse, a liability. Moths generally aestivate in caves with a temperature of approximately 9°C (Common 1954), so much higher temperatures may be disruptive. As the bogong moth is of immense cultural and ecological value—for example, 'perhaps the only native insect in Australia to have been accorded notoriety' (New 2007) with regards to its spectacular massing on public buildings—and it is also an essential food source for the endangered mountain pygmy possum, *Burramys parvus*, its absence from the alpine environment will be widely detrimental.

Tiling moths are predicted to retain moisture more effectively than isolated moths, and tiling is also predicted to be a response to cooler nights or cave conditions in alpine environments. By analysing the tiling response of moths under temperature and humidity extremes, one should be able to assess the impact of changing conditions on moth populations. Tiling response to temperature and humidity change is predicted to be conservative, in the sense of preserving both temperature and resources.

Method

Moths were captured overnight in light traps supplied by the NSW Parks and Wildlife Service. These light traps were placed in a boulder field by the road at Charlotte Pass where bogong moths had previously been identified, and were checked for two consecutive nights in mid-December. Moths captured each night were removed from the traps and were kept in dark containers with other moths at 25°C and ambient humidity.

Once sufficient specimens were collected (>110 was deemed sufficient), containers were briefly cooled to 4°C to enable non-traumatic moth handling, and captured moths of species other than *A. infusa* were released.

Moths destined for the humidity treatment were placed in rectangular clear plastic containers, in groups of 10 or 11. Solo enclosures were constructed for the humidity treatment: made of cardboard, fine mesh and clothes pegs, of a size designed to allow insertion of a single moth and moth movement without escape or actual flight.

The containers for the humidity treatment were filled with a layer of orange silica desiccant balls and were weighed. Moths were added (either 10 or 11 moths, depending on box and on number of moths available) and the container weighed again. Weight of solo enclosures with and without a moth was also determined, and then the moths were placed in a dark room (in order to control for the possibility of light/dark cues for tiling behaviours) at 25°C. Moth mass loss was recorded after 24 hours, for both groups and isolated moths. All mass loss was considered to be water loss, as other routes for mass loss (e.g. frass) were negligible. Moth mass loss was considered a more reliable method of measuring water transfer than silica mass gain, as net gain in silica mass could easily have been atmospheric, from the air within the container, or incomplete.

Moths destined for the temperature treatment were placed in rectangular clear plastic containers, randomly allocated to either hot or cold, with each treatment further containing four groups, comprising 3, 5, 8 or 10, and 13 moths. No replications were possible due to low number of available moths.

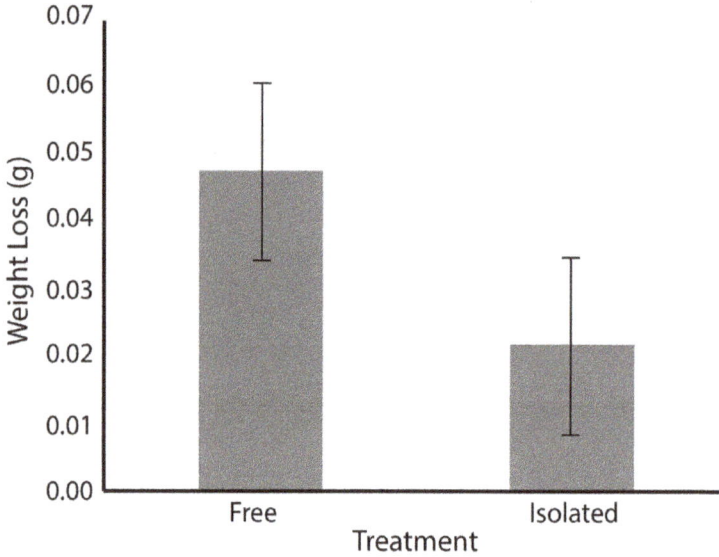

Figure 1: Weight loss (g) of free and isolated moths kept in low humidity (n=5).

Note: Error bars are standard deviation.

Source: Ming-Dao Chia.

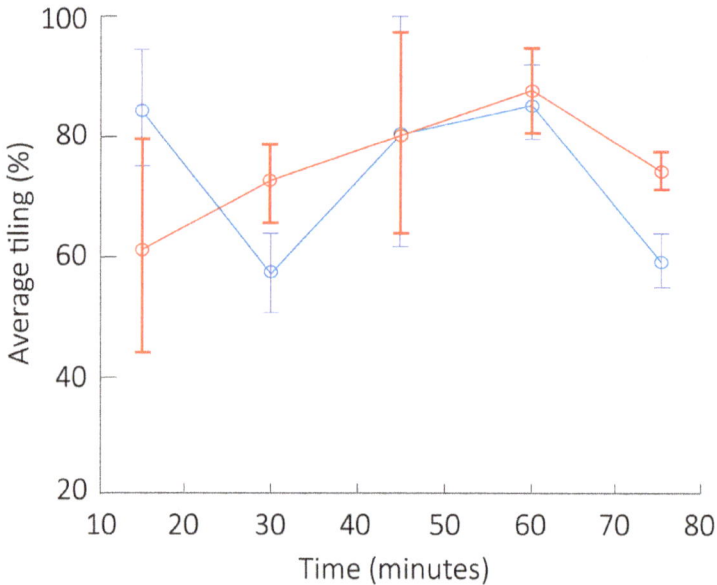

Figure 2: Moth tiling (as percentage) in hot (red) and cold (blue) treatments vs time in minutes.

Note: Error bars are standard deviation.

Source: Ming-Dao Chia.

Moths undergoing temperature treatment were kept at either 25°C (hot treatment) or 4°C (cold treatment) and had temperature and number tiling measured at 15-minute intervals for a total of 75 minutes. These temperatures were selected in order to provide extremes compared to the standard aestivation conditions of bogong moths, in caves ranging between 9°C and 14°C (Common 1954). Temperature was measured both of the ambient air and of the tiling moths. Temperature measurement was taken by a Fluke 568 infrared thermometer, aimed at a piece of cardboard attached to the box and then aimed at any groups of tiling moths. Number of moths tiling was calculated from photographs taken at the same time as the temperature measurements.

After 24 hours for the humidity moths and 75 minutes for the temperature moths, all specimens still alive were released into the boulder field from which they were initially captured.

Results

Data were collected on average moth water loss, calculated by subtracting the final weight of the container and silica without moths from the initial weight of the container with moths, and then dividing the result by the number of moths. Average moth weight loss was calculated (rather than total weight loss per box) in order to compare more accurately to weight loss in the isolated moths. There was no statistically significant relationship between group presence and weight loss, although there was a trend towards lower weight loss in isolated moths (Figure 1).

As moth tiling generally occurs at approximately 9°C (Common 1954), temperature extremes of 25°C and 4°C were selected. Unfortunately, measurement was restricted to a 75-minute time period, as moths in smaller-sized groups began to exhibit high mortality rates. Dead moths were not removed, and were considered as tiling/non-tiling based on their location at death. Mortality occurred only in the heat treatment groups, and then only in the heat treatment groups with five moths or fewer. Moth mortality is thus not accounted for in calculations of tiling. Over the given time period of 75 minutes, no statistically significant relationship between moth percentage tiling and temperature was demonstrated (Figure 2).

Discussion

The results show no definite statistical relationship, but do indicate that trends exist. Free moths lost a near-significant amount of body mass (as water weight) when compared to isolated moths. This was contrary to hypothesis, and requires further investigation. Tiling moths only showed a significant difference between temperature groups at the 75-minute mark, with moths tiling much more in warmer groups than in cold. It is important to consider that moths may have taken longer to tile than expected, and that the duration of observations was too short. Neither temperature is outside the natural range of moths, though it is outside their range of aestivation temperatures. This result corresponds to the hypothesis, considering that tiling moths may attempt to be conserving water. However, considering the water loss in tiling moths was greater, rather than less, increased tiling under warm conditions may serve a different purpose.

Replication trials were impossible to conduct due to the difficulty of sustainable moth collection, and temperature regulation relied on equipment not designed for the purpose. With these limitations taken into consideration, the hypothesis may require refinement rather than rejection.

While the existing literature, though limited, indicates a potential positive relationship between tiling percentage and survival of extreme temperature and dryness conditions (Common 1954), nothing in the current work indicates definitively any kind of relationship between tiling and moth survival under either set of conditions. The absence of statistical significance in any direction indicates that either the trigger for tiling behaviours is something else entirely, or that the method was flawed. Either way, there is no demonstrable causal relationship between either humidity levels and tiling or temperature and tiling. Given that animal behaviour is rarely simple, and rarely stems from a single cause, attributing the tiling behaviour to one cause, or even to two, seems to be a flawed response in and of itself. Possibly the lack of statistical significance was due to a failure to observe many of the other potential factors. As several standard measures of circadian rhythm (for example, eye pigmentation; Common 1954) indicate that the trigger for tiling behaviours is unrelated to day length, that is less likely to be a cause, but there are still several other potential factors left unaccounted for.

Tiling may be a social response, as indicated in some observations by Common (1954). Moths form aggregations around centres, although it is still unclear whether this is based on condition optimality or on some kind of herd behaviour. Without anthropomorphising too greatly, this 'social response' may also be defence against predation. *Agrotis infusa* already possesses false eyes, but a massing display may be perceived as a threat to predators, or perhaps the moths are merely finding safety in numbers.

Without rearing moths in the laboratory or testing temperatures in the field, it is impossible to be entirely sure that mortality and tiling patterns are not responses to procedure-based drama (handling, exposure to humans, etc.). The box transition process in particular was incredibly distressing to moths, with some requiring repeated heating to room temperature and then cooling to below handling temperature before they were transferred to their final container. While this would certainly explain mortality rates in handled moths, some moth trauma was unfortunately unavoidable.

Given the short time, low n-value and lack of replications, however, these results may indicate nothing more than that the method was flawed. The lack of replications make proper statistical analysis impossible, and the lack of any trend may be indicative of a lack of relationship, or it may be that non-trend results were simply anomalous and poorly accounted for.

The humidity treatment was hard on the moths, and exposing moths to extremely dehydrated conditions may have destroyed their mechanisms for coping with the mildly dehydrated conditions any humidity response to tiling may protect against. Future research should focus on providing humidity gradients closer to those found in the field, over longer periods of time, rather than relying on extremes to provide accurate data for conditions occurring in the middle.

The limited time made it impossible to determine whether or not mortality rates were in fact due to temperature or to tiling percentage. Running the temperature tests for longer would have given some indication of whether stabilisation would happen and a trend be visible in groups.

Even if tiling is effective, it might be so within such a limited range that the admittedly brute-force techniques used in this trial would have been insufficient to observe any trends. Rarely do alpine rock shelters in the shade reach 25°C, for example, so any warming effect of tiling may be of more use to moths at cold temperatures, and a completely different tiling response may have been seen in moths kept at a theoretical 18°C. A greater range of temperature tests would allow observation of any range of tiling behaviours, lending some support to the notion that tiling serves a thermal purpose.

Tiling may also serve jointly, creating higher humidity while also increasing temperature. Perhaps under natural conditions, ambient temperature and humidity are such that the moths run very little risk of heat-based mortality, while using tiling to maintain safe humidity conditions.

Examination of a greater range of temperatures, a greater and more stable range of humidity, using more field-like containers and specimens exposed to less trauma may still indicate some relationship between tiling and heat or humidity. Should that fail, examination of other social and environmental factors may prove key to explaining the peculiar tiling behaviour. CO_2 knockout procedures would allow handling to proceed more smoothly and would reduce moth trauma, making for more reliable results. The construction of moth solo enclosures of fine, breathable plastic mesh would be simpler, perhaps shaped something like a pit trap for initial entry, with the ingress then removed.

While very little can be done to prevent moth tiling (according to Common (1954), moths will tile on any available, sufficiently shaded surface once they reach their preferred aestivation sites) should it be detrimental to moth survival under warmer conditions, the provision of more appropriate locations for moth tiling may become a necessity should tiling behaviour act effectively to conserve moisture.

Isolated moths tended towards losing less weight, which may have been to do with mass and energy conservation as a stress response. If a moth is alone, excess movement could quite easily attract predators. Further, with no need to move towards tiling groups, the solo moths may have expended less energy. A trade-off between the energy needed to tile and the energy saved by tiling may exist, but have not been present under these conditions due to the humidity extremes to which moths were subjected.

Despite the prevalence of tiling behaviour in dry conditions, bogong moth tiling appears to have no significant effect on preventing water loss. Tiling is not demonstrably a response to either heat or cold, and in fact causes higher mortality rates at warm temperatures. The method used in this experiment was unsuitable for determination of any potential relationship between tiling behaviour and either heat or humidity.

Acknowledgements

We thank the Crowajingalong Scout leaders for invaluable brains, trust work and red wine, Michael Whitehead and Adrienne Nicotra for patience and expertise, Wes Keys and the lodge caretaker for support and supplies, ANU Department of Botany and Zoology for donation of resources, and Mel Shroder of NSW Parks and Wildlife Service for supplying moth traps.

References

Common IFB (1954) A study of the ecology of the adult bogong moth, *Agrotis infusa* (Boisd.) (Lepidoptera: Noctuidae), with special reference to its behaviour during migration and aestivation, *Australian Journal of Zoology* 2: 223–63. doi.org/10.1071/ZO9540223

New TR (2007) Politicians, poisons and moths: Ambiguity over the icon status of the Bogong moth (*Agrotis infusa*) (Noctuidae) in Australia. *Journal of Insect Conservation* 11: 219–20. doi.org/10.1007/s10841-007-9073-x

Welch HE (1963) *Amphimermis bogongae* sp.nov. and *Hexamermis* cavicola sp.nov. from the Australian bogong moth, *Agrotis infusa* (Boisd.), with a review of the genus *Amphimermis* Kaburaki & Imamura, 1932 (Nematoda: Mermithidae). *Parasitology* 53: 55–62. doi.org/10.1017/S003118200007253X

The effect of body mass on the mass specific metabolic rate of *Pseudemoia entrecasteauxii* and *Eulamprus quoyii*

Yvonne Yong, Ainsley Maurer, Caitlin McLeod, You Zhou

Abstract

Measuring metabolic rate is important to understand energy acquisition, transfer and expenditure within an organism. Mass specific metabolic rate is also known as metabolic intensity. The metabolic intensities of *Pseudemoia entrecasteauxii* and *Eulamprus quoyii* were measured within a closed system through the amount of oxygen depleted when at rest. There was found to be a negative correlation between the mass of the skink and the metabolic intensity, or the oxygen consumption per hour. The average mass of *P. entrecasteauxii* was found to be lower than *E. quoyii*, and *P. entrecasteauxii* was found to have a lower metabolic intensity than *E. quoyii*, which follows Kleiber's law. The difference in metabolic intensity may be influenced by the climate of the microhabitats that the skinks live in.

Introduction

All animals have chemical processes within them that keep them alive. Such processes include cell growth, brain and nerve function and respiration (Parry 1983). Metabolism is the term used to describe these processes of energy transactions that occur throughout the body and how the energy is transformed within the organism and the environment (Gillooly 2001).

The mass of an animal affects the rate of metabolism and this relationship is known as the mass specific metabolic rate, or metabolic intensity. This relationship is quantified by Kleiber's law (Kleiber 1947), which proposes that larger sized animals use less energy per cell to sustain their needs (Kalra et al. 2013). Kleiber's law states that metabolic rate of the organism increases at a rate proportional to its mass to the power of 0.75, which means that larger animals have a lower metabolic rate per unit mass compared to smaller animals. Thus, there is a negative correlation between the metabolic rate per unit mass, also known as metabolic intensity, and mass (Kleiber 1947). This means that smaller organisms have a higher metabolic rate per unit mass compared to larger organisms.

The measurement of metabolic intensity, and hence metabolic rate, is important to describe energy acquisition, transfer or expenditure within an organism. This can be done through measuring oxygen consumption (Weir 1949). Measuring oxygen consumption is a common method to measure metabolic rate. Within all vertebrates respiration is mainly aerobic, meaning they consume molecular oxygen for cellular respiration. The assumption is made that oxygen consumption is an estimate of the rate of energy metabolism (Schmidt-Neilsen 1990; Speakman 2013).

To measure the oxygen consumption, and hence the metabolic rate, a closed system is often used as the decline in oxygen is easily measureable. The organism is measured at rest to find out the standard metabolic rate (or basal metabolic rate in endotherms). Standard metabolic rate is what the minimum metabolic rate is to sustain the organism's life at the certain temperature.

Ectotherms have their standard metabolic rate dependent upon the surrounding temperature. Skinks are an example of an ectothermic animal. Skinks can be found in a wide range of habitats, including the Australian alpine region. Two such species that are found in the Australian alpine region are *Pseudemoia entrecasteauxii* (mountain log skink) and *Eulamprus quoyii* (eastern water skink). *Pseudemoia entrecasteauxii* can be found in a variety of preferably sunny habitats, such as on rocks and north-west aspects (Jellinek et al. 2004), whereas *E. quoyii* are commonly found nearby creeks or moist areas (Law and Bradley 1990). The unique microhabitats that these species live in reflect their optimal temperatures, at which their metabolic rates are at their highest. It would be assumed that *E. quoyii* has a lower optimal temperature based upon its microhabitat of near moist areas, as they would be cooler. Similarly, based upon the microhabitat of *P. entrecasteauxii*, it would be assumed that their optimal temperature is higher than that of *E. quoyii*.

The main aim of the experiment is to compare the metabolic intensities of the two skink species, *P. entrecasteauxii* and *E. quoyii*. This is explored by investigating the effect body size has upon the metabolic intensity as well as how the habitats of the skinks affect their optimal temperatures and therefore their metabolic intensities at a given temperature.

The predicted comparison of metabolic intensities of *P. entrecasteauxii* and *E. quoyii* is that *P. entrecasteauxii* should have a lower metabolic intensity compared to *E. quoyii*. This is because they are smaller organisms and their optimal temperature may be further from the ambient experimental temperature than that of *E. quoyii*.

Methods

The skinks were all caught around Rainbow Lake in Kosciuszko National Park on 9 and 11 December 2015. There were eight *Pseudemoia entrecasteauxii* and seven *Eulamprus quoyii*. The *P. entrecasteauxii* were caught on 9 December whereas the *E. quoyii* were caught on 11 December, both around 9 am to 12 pm. The weather on both days was sunny and slightly windy; however, it should be noted that 8 December had severe rain while 10 December had pleasant weather. The experiments were completed the day after capture and the skinks were placed back where they were caught on the day when experiments were completed. The skinks were caught by using mealworms to entice them within range of capture by hand. The skinks were housed separately in containers (~8 cm × 13 cm × 25 cm) with mulch, some water in a bottle cap and a cardboard tube for shelter. For the duration of the experiments, the skinks were not given any food.

The equipment used for measuring oxygen consumption was the Firesting Oxygen fibre-optic oxygen meter with Pyro-Science Oxygen Logger Software version >3.2. Four custom-built chambers were used in each trial and had Pyro-Science Contactless Fibre-Optic Oxygen Sensor Spots inside. Pyro-Science Basic Spot Adapters were attached on top of the spots using sticky tape and a needle was used to make a hole for the sensor. The chambers were calibrated to a stable amount of oxygen, which was as close to 21 per cent as possible, and the ambient temperature during experiments was kept at 17.5°C to 18.5°C through the use of a heat lamp. At night, the lamps were turned off to simulate the night environment and to keep the skinks less active.

The lizards were numbered in order of capture and the order in which they would be placed in the chambers was randomised using the random number generator implemented in Microsoft Excel. The chamber sizes used were 220 mL and 380 mL. The 220 mL chambers were used for *P. entrecasteauxii* and the 380 mL were predominantly used for the *E. quoyii*.

The lizards' oxygen consumption was measured at rest inside the chambers for 15 minutes. The measurement started when the skink was no longer breathing heavily, and if the oxygen percentage within the chamber lowered by 3 per cent or more within the 15-minute timeframe, the experiment would be stopped for that chamber due to a lack of oxygen. Furthermore, the skinks' behaviours were noted while in the chambers and during the transfer between the containers and chambers.

Three experiments were conducted for each lizard and their masses weighed. The data were averaged and the metabolic intensity calculated. The units used for metabolic intensity are mL of oxygen/gram/hour. The metabolic intensities (MI) of the skinks was calculated using the following formula:

$$MI = \frac{((Start\ O_2\% - Finish\ O_2\% \times 100) \times ((Chamber\ Volume - Skink\ Mass) \times 4))}{(Mass\ of\ Skink)}$$

Results

We found a negative correlation between the mass of the skink and the average oxygen consumption (Figure 1). *Pseudemoia entrecasteauxii* does not have a strong negative correlation between its mass and the oxygen consumption (Figure 1c), particularly when compared with *Eulamprus quoyii*, which has a stronger negative correlation between the mass and oxygen consumption. Furthermore, the average mass of *E. quoyii* is higher than that of *P. entrecasteauxii* (Figures 1a and 1b). When compared logarithmically, the correlation between the mass of *P. entrecasteauxii* and its metabolic intensity is not as strong as the correlation of mass and metabolic intensity of *E. quoyii* (Figure 1c). The strength of the correlations can be seen through the R-squared values (Figure 1c). *Pseudemoia entrecasteauxii* has an R-squared value of 0.7084 whereas *E. quoyii* has a value of 0.91823. A logarithmic comparison of the metabolic intensities between the two skinks shows that *E. quoyii* has a higher metabolic intensity than that of *P. entrecasteauxii*. In addition to this, the *P. entrecasteauxii* has a trend line lower than that of the *E. quoyii* (Figure 1c).

Skinks were observed to be heavily breathing and some were still moving around during the experimentation, which were often signs of stress. The skinks with a larger mass were observed to have a tendency to escape when the enclosures were opened.

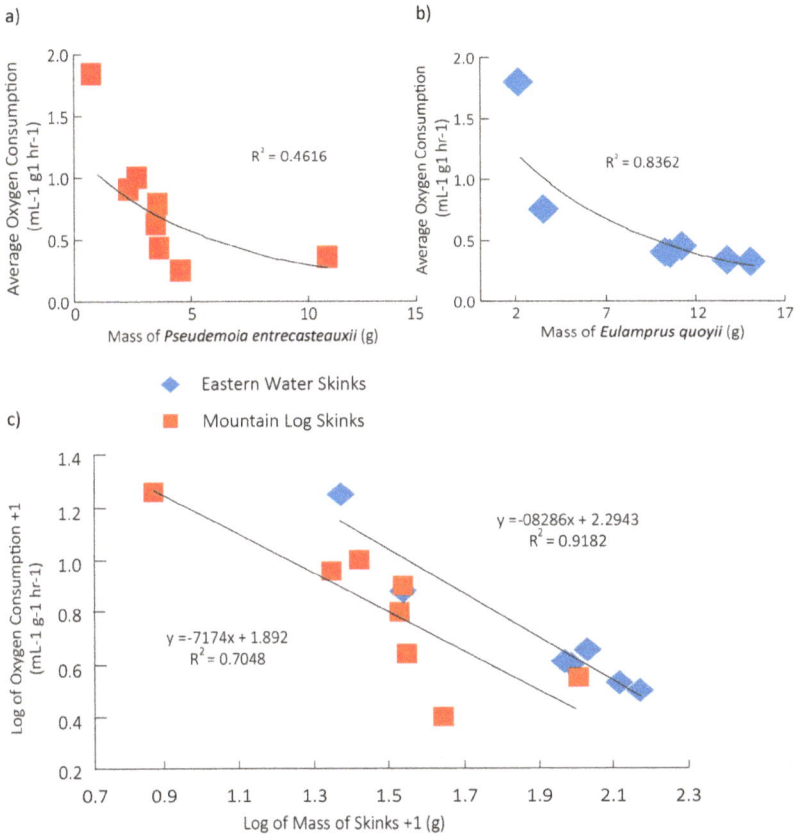

Figure 1: Average oxygen consumption of a) *Pseudemoia entrecasteauxii*, and b) *Eulamprus quoyii* as a function of body mass, c) log of the mass of the skinks +1 against the log of oxygen consumption +1 at 18°C.

Note: The blue diamond plots are for *E. quoyii* and the red square plots are *P. entrecasteauxii*. All the data points had 1 added to them to give positive values.

Source: Authors' data.

Discussion

The results show a trend for larger *P. entrecasteauxii* and *E. quoyii* to have a lower metabolic intensity than smaller-sized *P. entrecasteauxii* and *E. quoyii* (Figures 1a and 1b). This is seen through the negative correlation between the oxygen consumption and the mass of the skink. The comparison of both *P. entrecasteauxii* and *E. quoyii* on a logarithmic scale (Figure 1c) shows that *P. entrecasteauxii* has a lower metabolic intensity than *E. quoyii* at the ambient temperature of approximately 18°C. The average body size of *P. entrecasteauxii* can also be seen to be less than that of *E. quoyii* within Figure 1c. The results agree with Kleiber's law, which states that there should be a negative correlation between metabolic rate per unit against body size (Kleiber 1947). The lower metabolic intensity and the smaller average body size of *P. entrecasteauxii* agrees with Kleiber's law that states smaller body sizes correlates to a lower metabolic intensity (Kleiber 1947). Furthermore, the lower metabolic intensity of *P. entrecasteauxii* may suggest that they have a higher optimal temperature, reflected by their habitats of sunnier, warmer areas when compared to *E. quoyii*.

The results show a trend for larger *P. entrecasteauxii* and *E. quoyii* to have a lower metabolic intensity than smaller sized *P. entrecasteauxii* and *E. quoyii* (Figures 1a, 1b). This is seen through the negative correlation between the oxygen consumption and the mass of the skink. The comparison of both *P. entrecasteauxii* and *E. quoyii* on a logarithmic scale (Figure 1c) shows that *P. entrecasteauxii* has a lower metabolic intensity than *E. quoyii* at the ambient temperature of approximately 18 °C. The average body size of *P. entrecasteauxii* can also be seen to be less than that of *E. quoyii* within Figure 1c. The results agree with Kleiber's law, which states that there should be a negative correlation between metabolic rate per unit against body size (Kleiber 1947). The lower metabolic intensity and the smaller average body size of *P. entrecasteauxii* agrees with Kleiber's law that states smaller body sizes correlates to a lower metabolic intensity (Kleiber 1947). Furthermore, the lower metabolic intensity of *P. entrecasteauxii* may suggest that they have a higher optimal temperature, reflected by their habitats of sunnier, warmer areas when compared to *E. quoyii*.

There were many limitations encountered when gathering and processing the field data. These include time constraints when allowing the skinks to rest before chambers were closed to begin the timing of oxygen depletion.

Chambers were closed when the skinks were still heavily breathing to reduce the time needed for each trial. This could have led to skewed results because the skinks are inhaling more oxygen than if they were at rest.

The many assumptions required were also a limitation when processing the data. These assumptions include that the skinks were fasting, were not gravid, were of good health, were all similar ages, that the ambient temperature was consistently kept between 17.5°C and 18.5°C and that the same amount of oxygen results in the same amount of ATP. Obvious amendments to the method of the experiment would be doing the experiments in better-controlled areas of temperature and have consistent health of the skinks. The stage of fasting within the skinks could be controlled by having the skinks captive for a longer period of time, ensuring that the skinks are fed the same diet for a set amount of time before fasting them.

There have been no previous measurements of the metabolic intensities measured through oxygen depletion using a closed system for *P. entrecasteauxii* and *E. quoyii*. Further research to explore the metabolic intensities of the skinks includes investigating the microhabitats in which *P. entrecasteauxii* and *E. quoyii* are found, which can assist in determining the optimal temperatures for species based upon their habitats. Similarly, a study of the skinks' enzymatic activities can be done to further investigate the optimal temperatures within *P. entrecasteauxii* and *E. quoyii*.

Acknowledgements

We acknowledge the Research School of Biology of The Australian National University and especially Adrienne Nicotra for making the BIOL2203 course a reality, Phillipa Beale for guiding the group through the experiment, William Foley for guidance through the write up, and rangers and staff of Kosciuszko National Park for permission to collect organisms and experiment within the park.

References

Gillooly JF (2001) Effects of size and temperature on metabolic rate. *Science* 293: 2248–51. doi.org/10.1126/science.1061967

Janzen, DH (1967) Why mountain passes are higher in the tropics. *The American Naturalist* 101(919): 233–49.

Jellinek S, Driscoll DA, Kirkpatrick JB (2004) Environmental and vegetation variables have a greater influence than habitat fragmentation in structuring lizard communities in remnant urban bushland. *Austral Ecology* 29: 294–304. doi.org/10.1111/j.1442-9993.2004.01366.x

Kalra S, Shah S, Sahay R (2013) Kleiber's law and the A_1 chieve study. *Indian Journal of Endocrinology and Metabolism* 17: 397. doi.org/10.4103/2230-8210.122039

Kleiber M (1947) Body size and metabolic rate. *Physiological Reviews* 27: 511–41.

Law BS, Bradley RA (1990) Habitat use and basking site selection in the water skink, *Eulamprus quoyii*. *Journal of Herpetology* 24: 235–40. doi.org/10.2307/1564388

Parry GD (1983) The influence of the cost of growth on ectotherm metabolism. *Journal of Theoretical Biology* 101: 453–77. doi.org/10.1016/0022-5193(83)90150-9

Schmidt-Nielsen K (1990) *Animal Physiology: Adaptation and Environment*, 4th edn. Cambridge University Press, Cambridge.

Speakman JR (2013) Measuring energy metabolism in the mouse – theoretical, practical, and analytical considerations. *Frontiers in Physiology* 4. doi.org/10.3389/fphys.2013.00034

Weir JBdeV (1949) New methods for calculating metabolic rate with special reference to protein metabolism. *The Journal of Physiology* 109: 1–9. doi.org/10.1113/jphysiol.1949.sp004363

Predicting ectotherm vulnerability to climate warming: Comparing preferred and actual body temperatures in *Pseudemoia entrecasteauxii*

Julia Hammer, Hannah Zurcher, Christine Mauger, Chen Liang, Ming-Dao Chia, Tess Walsh-Rossi, Angela Stoddard, Cameron McArthur, Sarah Stock

Abstract

The performance of ectotherms can be related to temperature: performance increases with temperature to a certain optimum, above which it quickly drops. Climate warming is a threat to tropical ectotherms, whose thermal optima are relatively low. Temperate and alpine ectotherms, in contrast, have higher thermal optima, and therefore may benefit from rising temperatures. The preferred body temperatures (T_p) and field body temperatures (T_b) of viviparous Australian alpine skink, *Pseudemoia entrecasteauxii* (Duméril and Bibron), were compared. Results showed that T_p was lower than T_b in early summer, implying that short-term warming could enhance performance. These thermal performance parameters are one of many factors that require consideration when predicting the vulnerability of ectotherms to climate warming.

Introduction

Climate warming presents a permanent and pervasive challenge to organisms worldwide (IPCC 2015; Parmesan 2006). Consequently, biologists are interested in developing conservation and management strategies that prioritise vulnerable species (Williams et al. 2008).

Ectotherms are arguably the most sensitive to warming, since performance depends on body temperature, and body temperature depends on the environment (Angilletta et al. 2002). Ectotherms operate at body temperatures (T_b) between specific critical thermal minima (CT_{min}) and maxima (CT_{max}; Huey et al. 2012). Within those limits, performance is highest at a certain optimal temperature (T_o), but most ectotherms thermoregulate to a preferred body temperature (T_p) just below this optimum. When $T_b < CT_{min}$, performance gradually lessens, but when $T_b > CT_{max}$, performance rapidly drops to lethality. The extent to which an ectotherm is vulnerable to climate warming depends on whether it is currently achieving or exceeding T_p ($T_b \geq T_p$), or not achieving T_p ($T_b < T_p$): if the former were true, climate warming would push T_b towards CT_{max}, reducing performance and restricting activity, but if the latter were true, warming would bring T_b closer to T_p, enhancing performance and enabling extended activity (Huey et al. 2012).

Highland ectotherms are exposed to cooler, more variable temperatures than tropical ectotherms, thus selection should favour broader thermal tolerance in alpine areas compared to the tropics (Janzen 1967). Indeed, recent analyses suggest that tropical lizards are more in danger of extinction due to climate warming (Sinervo et al. 2010) than temperate and alpine lizards (Aguado and Braña 2014; Caldwell et al. 2015). Nevertheless, thermal performance curves vary between species, and whether *all* temperate lizards benefit from short-term climate warming remains unclear (reviewed in Clusella-Trullas and Chown 2014). The present study compared T_p and T_b in the Australian alpine skink *Pseudemoia entrecasteauxii* (Duméril and Bibron 1839) to evaluate its vulnerability to future climate warming.

Pseudemoia entrecasteauxii are found in south-east Australian forests and grasslands (Cogger 2014). They are active roughly between December and March (Pengilley 1972 cited in Stapley 2006), during which they progress through viviparous reproductive stages—autumn spermatogenesis and mating (sperm storage over winter hibernation) and spring/summer births (Murphy et al. 2006). Both sexes must monitor food consumption to meet the energy demands of each stage, and gravid females must carefully thermoregulate to maximise performance (high T_o) *and* embryonic survival (low T_o; Beuchat and Ellner 1987). If climate warming results in $T_b > CT_{max}$ during spring/summer, fecundity would suffer; however, if warming pushes a low T_b closer to T_p, then extended foraging time may increase fecundity. Given their complex ecology, and that their active months are the hottest of the year (BOM 2015), predicted to get hotter (Worboys and Good 2011), this species provides a compelling system for examining climate warming impacts on ectotherms.

The present study aimed to test whether alpine ectotherms would benefit or not from future climate warming during one part of the year. It was hypothesised that *P. entrecasteauxii* would not be achieving T_p ($T_p > T_b$) in early summer, and therefore benefit from future warming.

Methods

Lizard collection and husbandry

All materials and equipment were provided by The Australian National University's Research School of Biology. Twelve *P. entrecasteauxii* (three males, nine females) were collected by hand or with mealworm baits from the Rainbow Lake walking trail in Kosciuszko National Park, mid-December 2015. In the laboratory, lizards were transferred into individual plastic tubs with mesh windows. Tub floors were covered in soil from the capture site. Dishes of water and toilet rolls for shelter were placed inside, but no food was provided. Bright indoor lights were kept on during daylight hours, and temperature was kept at ~20°C. The lizards were handled by another research party for two days following capture.

Measuring T_p and field T_b

Preferred body temperature was determined using a thermal gradient. A 150 cm × 30 cm × 15 cm Perspex chamber with two runs (separated by a Perspex wall) and a removable lid were set up in a quiet room. A linear temperature range from ~7°C to 47°C was achieved inside the chamber with a heat lamp at one end. Lizards were placed in the thermal chamber in pairs (one per run) and left alone for five minutes to achieve T_p. Lizard body temperature was measured by pointing an infrared thermal gun (FLUKE 568 or Digitech QM7215) on the dorsal side between the shoulders. This was repeated at midday the following day. Body and substrate temperatures in the field were recorded by pointing an infrared thermal gun at log- or rock-basking *P. entrecasteauxii* along the Rainbow Lake walking trail, also around midday.

Results

Mean T_p was found to be significantly higher than mean field T_b using a Student's *t*-test (Figure 1; $P<0.001$). This suggests that *P. entrecasteauxii* was not achieving T_p in the field at the time of collection (early summer). Air temperatures (between 23°C and 26°C) and substrate temperatures (logs and rocks; ~28°C) were fairly consistent with T_b. Conditions during collection were sunny, partly cloudy and slightly windy.

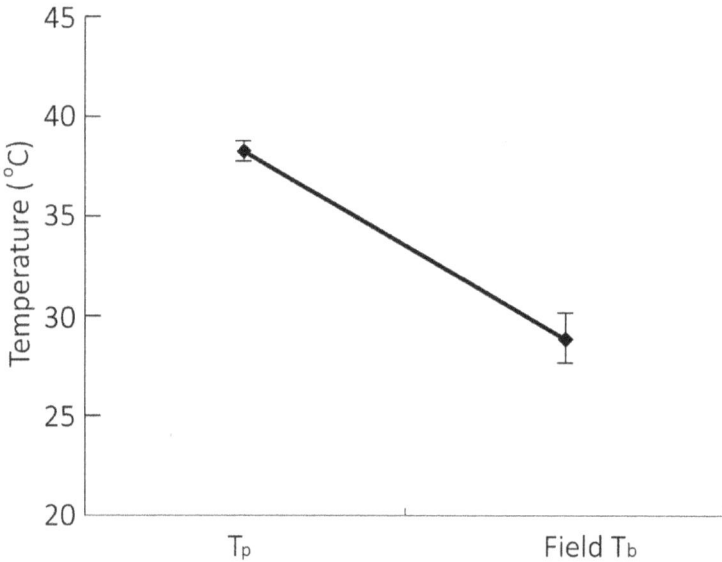

Figure 1: Mean T_p (n=12) and field T_b (n=13) for *P. entrecasteauxii*, calculated from thermal gradient and field data.
Note: Error bars represent standard error from the mean.
Source: Authors' data.

Discussion

Pseudemoia entrecasteauxii achieved a lower T_b than T_p for early summer, implying that a climate warming-induced increase in T_b would not be detrimental to reproduction, but rather enhance performance and extend active hours for foraging. This supports the hypothesis that alpine lizards may benefit from short-term warming, and agrees with past research (Aguado and Braña 2014; Caldwell et al. 2015). Nevertheless, several limitations must be noted.

Sample size was small, and predominantly female. Beal et al. (2014) found that body temperatures of *Sceloporous jarrovii* varied between sexes: it is possible that the data are only representative of female lizards.

Lizards were left in the thermal gradient for only five minutes, and their body temperatures were measured in a relatively intrusive manner. Stapley (2006) allowed her lizards one month to familiarise themselves with their thermal gradient, and used a thermal camera to collect temperature data. It is possible that the lizards in this study were agitated by handling and human presence such that achieving T_p was not prioritised.

The current data only describe a few days within a season—it is unknown whether *P. entrecasteauxii* achieve T_p in autumn and hence also unknown whether climate warming is a year-long benefit.

It is therefore recommended that further research on this topic involve a large and diverse sample size, a thermal gradient in which lizards can comfortably and naturally achieve T_p with minimal interference, and repeated collections and measurements throughout the active season.

Comparing T_p and field T_b, while useful, is only one of many factors that contribute to ectotherm climate change vulnerability. Others include the extent to which behavioural thermoregulation and microclimate heterogeneity can buffer warming, the plasticity and adaptive potential of thermal performance parameters (i.e. T_p and CT_{max}), and the impact of biotic and abiotic forces (besides temperature) that result from climate change (Clusella-Trullas and Chown 2014; Huey et al. 2012). For example, Kearney et al. (2009) predict that an increase in air temperature in Australian temperate zones is easily managed by shuttling between sun and shade. Whether *P. entrecasteauxii* is capable of shuttling through future climate warming is worth investigation. One suggestion is to survey the available temperatures for *P. entrecasteauxii* with absorptive lizard-shaped objects (see Sunday et al. 2014), and model how those temperatures might change with climate warming. According to Huey et al. (2003), ectotherms are incapable of adapting to climate warming due to physiological constraints: thermoregulation allows ectotherms to avoid unwanted temperatures and therefore weakens selection pressures for higher thermal tolerances. Perhaps a close examination of phylogenies (see Clusella-Trullas and Chown 2014) will reveal which lizard families are most at risk.

In conclusion, it is possible that highland ectotherms, like *P. entrecasteauxii*, will benefit from short-term climate warming; however, accurately predicting species' responses to climate change requires a broader, more comprehensive approach—a direction for future research.

Acknowledgements

The authors thank Craig Moritz and Scott Keogh for their guidance, Jack Keogh for lizard-wrangling, the South Alpine Ski Club at Charlotte Pass for their hospitality, and the ANU Research School of Biology for providing all materials and equipment.

References

Aguado S, Braña F (2014) Thermoregulation in a cold-adapted species (Cyren's Rock Lizard, *Iberolacerta cyreni*): Influence of thermal environment and associated costs. *Canadian Journal of Zoology* 92: 955–64. doi.org/10.1139/cjz-2014-0096

Angilletta MJ, Niewiarowski PH, Navas CA (2002) The evolution of thermal physiology in ectotherms. *Journal of Thermal Biology* 27: 249–68. doi.org/10.1016/S0306-4565(01)00094-8

Beal MS, Lattanzio MS, Miles DB (2014) Differences in thermal physiology of adult Yarrow's spiny lizards (*Sceloporus jarrovii*) in relation to sex and body size. *Ecology and Evolution* 4: 4220–9. doi.org/10.1002/ece3.1297

BOM (Bureau of Meteorology) (2015) *Climate statistics for Australian locations*, viewed 18 January 2015, www.bom.gov.au/climate/averages/tables/cw_071003.shtml

Buechat CA, Ellner S (1987) A quantitative test of life history theory: Thermoregulation by a viviparous lizard. *Ecological Monographs* 51: 45–60. doi.org/10.2307/1942638

Caldwell AJ, While GM, Beeton NJ, Wapstra E (2015) Potential for thermal tolerance to mediate climate change effects on three members of a cool temperate lizard genus, *Niveoscincus*. *Journal of Thermal Biology* 52: 14–23. doi.org/10.1016/j.jtherbio.2015.05.002

Clusella-Trullas S, Chown SL (2014) Lizard thermal trait variation at multiplescales: A review. *Journal of Comparative Physiology* 184: 5–21. doi.org/10.1007/s00360-013-0776-x

Cogger H 2014, *Reptiles and Amphibians of Australia*, 7th edn. CSIRO Publishing, Clayton.

Huey RB, Hertz PE, Sinervo B (2003) Behavioural drive versus behavioural inertia in evolution: A null model approach. *The American Naturalist* 161: 357–66. doi.org/10.1086/346135

Huey RB, Kearney MR, Krockenberger A, Holtum JAM, Jess M, Williams SE (2012) Predicting organismal vulnerability to climate warming: Roles of behaviour, physiology and adaptation. *Philosophical Transactions of the Royal Society B* 367: 1665–79. doi.org/10.1098/rstb.2012.0005

IPCC (2015) Summary for Policymakers. In *Climate Change 2014: Synthesis Report*, IPCC, viewed 17 January 2016, www.ipcc.ch/pdf/assessment-report/ar5/syr/SYR_AR5_FINAL_full_wcover.pdf

Kearney M, Shine R, Porter WP (2009) The potential for behavioural thermoregulation to buffer 'cold-blooded' animals against climate warming. *Proceedings of the National Academy of Sciences of the United States of America* 106: 3835–40. doi.org/10.1073/pnas.0808913106

Murphy K, Hudson S, Shea G (2006) Reproductive seasonality of three cold-temperate viviparous skinks from southeastern Australia. *Journal of Herpetology* 40, 454–64. doi.org/10.1670/0022-1511(2006)40[454:RSOTCV]2.0.CO;2

Parmesan C (2006) Ecological and evolutionary responses to recent climate change. *Annual Review of Ecology, Evolution, and Systematics* 37: 637–69. doi.org/10.1146/annurev.ecolsys.37.091305.110100

Sinervo B, Méndez-de-la-Cruz F, Miles DB, Heulin B, Bastiaans E, Cruz MVS, Lara-Resendiz R, Martínez-Méndez N, Calderón-Espinosa M, De la Riva IJ, Sepulveda PV, Rocha CFD, Ibargüengoytía N, Puntriano CA, Massot M, Lepetz V, Oksanen TA, Chapple DG, Bauer AM, Branch WR, Clobert J, Sites JW (2010) Erosion of lizard diversity by climate change and altered thermal niches. *Science* 328: 894–9. doi.org/10.1126/science.1184695

Stapley J (2006) Individual variation in preferred body temperature covaries with social behaviours and colour in male lizards. *Journal of Thermal Biology* 31: 362–9. doi.org/10.1016/j.jtherbio.2006.01.008

Sunday JM, Bates AE, Kearney MR, Colwell RK, Dulvy NK, Longino JT, Huey RB (2014) Thermal-safety margins and the necessity of thermoregulatory behaviour across latitude and elevation. *Proceedings of the National Academy of Sciences of the United States of America* 111: 5610–15. doi.org/10.1073/pnas.1316145111

Williams SE, Shoo LP, Isaac JL, Hoffmann AA, Langham G (2008) Towards an integrated framework for assessing the vulnerability of species to climate change. *PLoS Biology* 6: 2621–6. doi.org/10.1371/journal.pbio.0060325

Worboys GL, Good RB (2011) *Caring for our Australian Alps Catchments: Summary Report for Policy Makers*. Department of Climate Change and Energy Efficiency, Canberra.

Sprint speed capacity of two alpine skink species, *Eulamprus kosciuskoi* and *Pseudemoia entrecasteauxii*

Isabella Robinson, Bronte Sinclair, Holly Sargent, Xiaoyun Li

Abstract

As global average temperatures continue to rise as a result of climate change, it is increasingly important to understand how some of the most vulnerable environments may be affected. The alpine environment and specialised biota of the Kosciuszko National Park are strongly influenced by abiotic factors such as temperature. Lizard performance in particular is closely related to temperature change. This study looks at the sprint speed capacity of two alpine skink species, *Eulamprus kosciuskoi* and *Pseudemoia entrecasteauxii,* as an indicator of fitness. Lizards were collected from two sites at Rainbow Lake and Charlotte Creek. These were raced over a 1 m distance and their sprint speeds were recorded at 25 cm intervals. Trials were conducted at room and elevated temperatures, and sprint times were compared between trials, species and sex and, for females, between gravid and non-gravid individuals. It was found that fitness, as measured through sprint speed, was greater at an elevated temperature for both species, and that *E. kosciuskoi* were significantly faster than *P. entrecasteauxii.* No significant differences were found between sexes or gravid and non-gravid individuals. It is possible that the lizard species studied would benefit from increased sprint performance linked to increased average temperatures; however, if temperatures rise above the skinks' physiological optima, it may have an extremely detrimental effect on all aspects of the lizards' biology.

Introduction

In the past hundred years, global average temperature has risen by 0.6 °C and is predicted to rise an additional 0.7°C by the year 2050 (Root et al. 2003; Pickering et al. 2004). This temperature change is already altering sensitive ecosystems and affecting the organisms within them, with a meta-analysis giving an average range shift of 6.1 km towards the poles and a 6.1 m shift upwards in altitude per decade (Pickering et al. 2004; Wyborn 2009). The commencement of seasonal spring events has been shown to be occurring on average 5.1 days earlier per decade in some species (Root et al. 2003). With significant changes already shown to be occurring (Root et al. 2003; Pickering et al. 2004; Wyborn 2009), it is increasingly important to understand the potential effects of climate change on the environments most severely at risk and their biota.

The alpine environment of the Kosciuszko National Park (KNP) is especially vulnerable to climate change, and its effects are already discernable in the changing snow and fire patterns (Pickering et al. 2004; Wyborn 2009). The functionality and differentiation of alpine environments is largely governed by temperature and other abiotic factors such as precipitation, especially that of snow (Pickering et al. 2004). Along with the forecasted rise in temperature, a dramatic reduction in snow cover and duration of snow coverage is expected. These changed conditions will have a considerable impact on the density and diversity of the region's specialised biota, as rising temperatures and reduction of snow coverage compress already narrow thermal environments and lowland species shift upward in altitude (Pickering et al. 2004). To best understand and predict changes to KNP's flora and fauna, it is essential to investigate the tolerances species might have to changed conditions and rising temperatures.

Understanding the thermal tolerances of KNP fauna will give an indication of the resilience of the species and help predict possible changes in distribution as temperatures increase. Reptiles, as ectotherms, are particularly sensitive to temperature and their fitness would be directly affected by changes to the annual means. Sprint speed capacity is a valuable indicator of species fitness. Faster individuals are more likely to avoid predation by fleeing and have an advantage as predators themselves. Greater speed has also been linked to social dominance and mating success (Beal et al. 2014). Maximal sprint performance has been linked to optimal temperature, and it has been shown in a number of lizard species that 'hotter is better' (i.e. they perform better at higher temperatures) (Zamora-Camacho et al. 2015). This has been shown to be equally true of warm- and cold-adapted species (Van Damme and Vanhooydonck 2001). However, other factors also have an influence on sprint capacity. The size and gravid status may all influence the lizard's speed and response to increased temperatures. Smaller and gravid lizards expected to be slower (Beal et al. 2014), though gravid individuals may have a higher sprint capacity at elevated temperatures due to having a higher thermal preference that benefits embryogenesis (Schwarzkopf and Shine 1991; Clusella-Trullas and Chown 2014).

We investigated the sprint capacities of two skink species found within KNP, *Eulamprus kosciuskoi* and *Pseudemoia entrecasteauxii*, at current and elevated temperature conditions, to see how rising temperatures within the park would affect the species. We predicted that both species would display greater sprint capacity with the increased temperature, but the smaller species, *P. entrecasteauxii*, would have a lower peak performance and that the gravid females of both would be significantly slower.

Methods

Study site

Kosciuszko National Park covers 690,411 hectares of Australian wilderness, making it one of the largest reserves in the country (Wyborn 2009). The park encompasses a large variety of biomes with distinct floral and faunal compositions (Pickering et al. 2004). The skink collection sites were Rainbow Lake, an artificially created lake off the Kosciuszko Road, and Charlotte Creek, which passes through the Charlotte Pass Snow Resort at 1,765 m.

Collection and husbandry

Twelve *P. entrecasteauxii* skinks were collected at the Rainbow Lake site using a baiting technique. Fishing rods were baited using mealworms tied to dental floss and were used to lure the lizards into the open and distract them while they were captured. Lizards were place in individually numbered fabric bags. Sixteen *kosciuskoi* skinks were collected using a noosing technique. Dental floss nooses were attached to the ends of fishing rods. These were slipped over the heads of the lizard and pulled tight, allowing for the lizard's capture. Captured lizards were placed in individually numbered fabric bags. Lizards were housed in individual, numbered plastic containers, containing a woodchip substrate, cardboard tube and a plastic drink lid containing water.

Measurements

The lizards were measured and their sex determined. Snout-to-vent and tail lengths were taken by stretching the lizard out against a clear ruler, and head length and width were measured with calipers. Weight was measured by placing lizards in a beaker on an electronic scale.

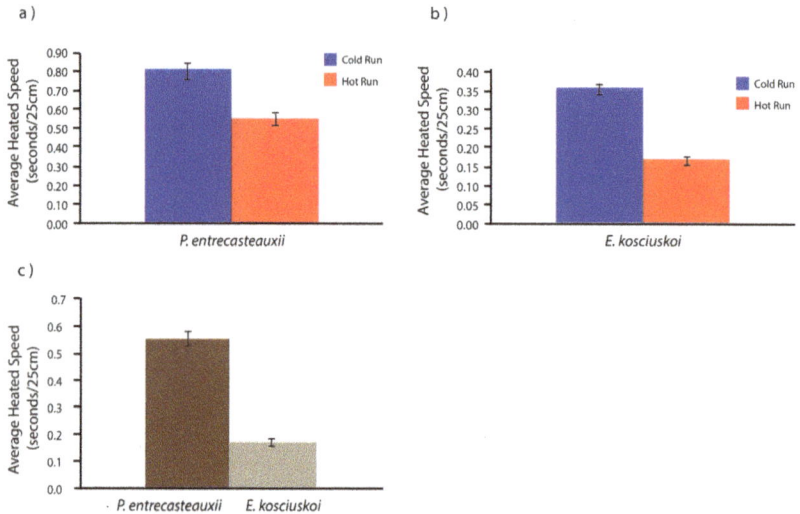

Figure 1: a) Comparison of *P. entrecasteauxii* speed between room temperature and heated runs, lizards were significantly faster after heating ($P=0.002$), b) Comparison of *E. kosciuskoi* speed between room temperature and heated runs, lizards were significantly faster after heating ($P<0.001$), c) Comparison of speed between species of lizard, *E. kosciuskoi* significantly faster than *P. entrecasteauxii* ($P<0.001$).
Source: Authors' data.

Lizard sprint speed was measured using a photocell-type racetrack. Lizards were released before the starting line and coaxed in the desired direction by a person running a broad paintbrush behind the lizard at a consistent speed. The racetrack terminated at the end of the table, where the lizard would be caught in a bucket. Lizard speed was recorded every 25 cm over the 1 m track and the fastest time was recorded. Each lizard was raced three times consecutively and two trials were conducted. One trial was conducted at room temperature and the second after the lizards were heated in paper bags for 1 hour at 32°C in an oven.

All lizards were released at the original sites of collection at the conclusion of the final trial.

Results

It was found that sprint performance in both *P. entrecasteauxii* and *E. kosciuskoi* was significantly improved after heating (*P*=0.002 and *P*<0.001 respectively), with *E. kosciuskoi* showing a greater increase in speed. *E. kosciuskoi* was shown to be significantly faster than *P. entrecasteauxii* (*P*<0.001). Differences in the average speeds of the female, male and gravid individuals were observed for both species, though were not found to be statistically significant (*P*=0.408 for *P. entrecasteauxii* and *P*=0.285 for *E. kosciuskoi*).

Discussion

An increase in the skink's body temperature resulted in an increase of average sprint capacity for both species. The overall average speed was greater for *E. kosciuskoi* and no significant difference was found between sexes or gravid and non-gravid individuals.

Global temperatures have risen 0.6°C in the last century and are set to rise an additional 0.7 °C over the next 50 years. This will have a significant effect on the alpine environment of the KNP. As temperatures increase, the thermal tolerances of the region's biota become increasingly important (Root et al. 2003; Pickering et al. 2004). The fitness of reptile species is closely correlated with temperature, and sprint speed capacity may be used as an indicator for lizard fitness (Beal et al. 2014). For the two lizard species studied fitness as measured through sprint speed was shown to increase at an elevated temperature.

For both species, sprint speed increased significantly after the lizards were heated, with *E. kosciuskoi* run times reduced by more than half (Figures 1a and 1b). This is consistent with the idea that 'hotter is better', as shown in previous studies on other lizard species (Zamora-Camacho et al. 2015). *E. kosciuskoi* lizards were significantly faster than *P. entrecasteauxii* lizards (Figure 1c). This is most likely due to their smaller size (Beal et al. 2014). Between the sprint capacities of the male, female and gravid individuals, no significant difference was discerned for either species, which is also consistent with the Beal study (Figures 2a and 2b), though it suggests the higher thermal preference of gravid females (Schwarzkopf and Shine 1991) does not result in a significant increase in speed.

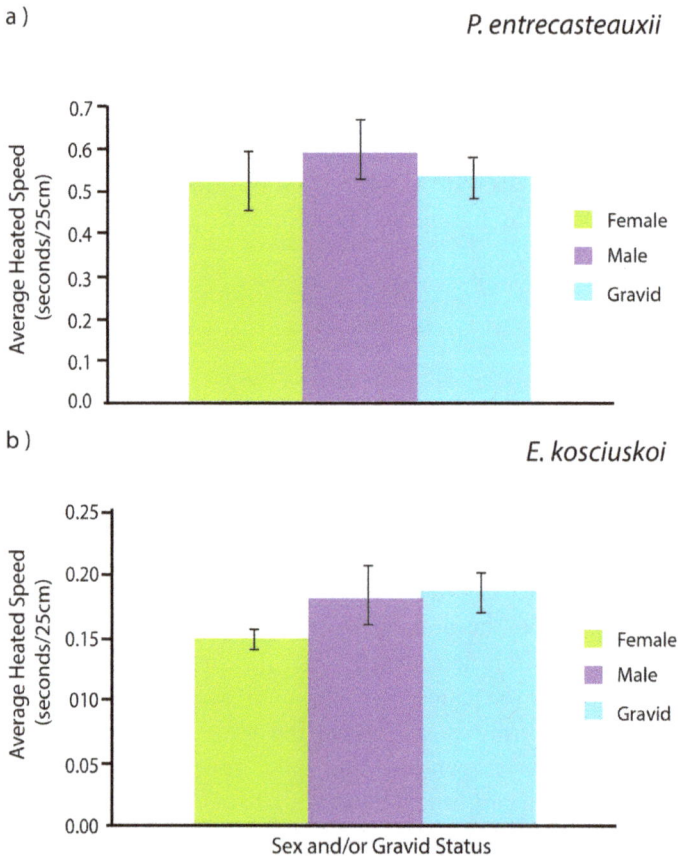

a)

P. entrecasteauxii

b)

E. kosciuskoi

Sex and/or Gravid Status

Figure 2: Comparison of heated run speed between male, female and gravid individuals of a) *P. entrecasteauxii* and b) *E. kosciuskoi*. No significant differences were observed between the groups.

Source: Authors' data.

Though the difference in sprint speed between room temperature and heated runs and between species were statistically significant, a large amount of variation existed between lizards within a trial and even within the run times of individual skinks. This may be due to a number of factors including inconsistency in the pace of prompting with the paintbrush and fatigue in the lizards that escaped from the racetrack, ran back toward the starting line or stopped and were required to run the track more than three times to obtain viable readings. An enclosed track and automated prompt may help reduce these sources of error. The ratio of sex and gravid status was also highly skewed towards females and gravid females in particular, as females tend to bask more, making them more likely to have been collected. This may have affected the comparisons in Figures 2a and 2b. It is also possible that the collected skinks were slower individuals, more susceptible to capture. A larger sample size, with a more even distribution of sex and gravid status would likely provide a more representative sample.

With fitness increasing at an increased temperature, it is possible that both lizard species will benefit from rising temperatures in the KNP region regardless of sex or gravid status. It may allow them to more easily escape predators, hunt more effectively and extend their ranges to higher altitudes as they warm. However, trials were run at only two temperatures and further trials at various temperatures would need to be conducted to determine the lizard species' optimal temperature and determine the effects of above optimal temperatures on lizard fitness. If temperatures continue to increase above the thermal preferences and tolerances of the skink species, it is likely that their physiology will be negatively impacted.

Acknowledgements

We thank Scott Keogh for his contribution to conducting the skink speed trials.

References

Beal, MS, Lattanzio MS, Miles DB (2014) Differences in the thermal physiology of adult Yarrow's spiny lizards (*Sceloporus jarrovii*) in relation to sex and body size. *Ecology and Evolution* 4: 4220–9.

Clusella-Trullas S, Chown SL (2014) Lizard thermal trait variation at multiple scales: A review. *Journal of Comparative Physiology B: Biochemical, Systemic, and Environmental Physiology* 184: 5–21.

Huey, RB, Schneider W, Erie GL, Stevenson RD (1981) A field-portable racetrack and timer for measuring acceleration and speed of small cursorial animals. *Experientia* 37: 1356–7.

Pickering, CG, Good R, Green RK (2004) *Potential Effects of Global Warming on the Biota of the Australian Alps: A Report for the Australian Greenhouse Office*. Australian Greenhouse Office, Canberra, ACT.

Root, TL, Price JT, Hall KR, Schneider SH, Rosenzweig C, Pounds JA (2003) Fingerprints of global warming on wild animals and plants. *Nature* 421: 57–60.

Schwarzkopf L, Shine R (1991) Thermal biology of reproduction in viviparous skinks, *Eulamprus tympanum*: Why do gravid females bask more? *Oecologia* 88: 562–9.

Van Damme R, Vanhooydonck B (2001) Origins of interspecific variation in lizard sprint capacity. *Functional Ecology* 15: 186–202.

Wyborn C (2009) Managing change or changing management: Climate change and human use in Kosciuszko National Park. *Australasian Journal of Environmental Management* 16: 208–17.

Zamora-Camacho, FJ, Rubiño-Hispán MV, Reguera S, Moreno-Rueda G (2015) Thermal dependence of sprint performance in the lizard *Psammodromus algirus* along a 2200-meter elevational gradient: Cold-habitat lizards do not perform better at low temperatures. *Journal of Thermal Biology* 52: 90–6.

Behavioural thermoregulation of alpine birds in response to low temperature in early summer

You Zhou, Ainsley Maurer, Tanja Cobden, Yvonne Yong,
Ray Zhang, Gregory Gauthier-Coles, Caitlin McLeod,
Cynthia Turnbull

Abstract

Thermoregulation in animals can be achieved physiologically and behaviourally. Alpine birds can obtain significant radiation benefits from the sun and to help conserve heat they adjust their behaviours such as perching. The aim of this study was to explore the preferences of alpine birds for temperature and illumination of perching sites. We measured the temperature and light intensity of both potential sites and actual sites where birds perched, and compared the distributions of temperature and illumination of these on cold and warm days among species. Results showed a non-random selection of temperatures on cold days and a slightly higher light preference on warm days. Among species, only the little raven (*Corvus mellori*) showed a preference for warmer and lighter perching sites, while other species either showed no predilection or perched in warmer places due to other factors of habitat selection. When perching, behaviours like sitting (hiding feet) and puffing were more frequent on cold days and no birds chose a shady site when the temperature was low. The influence of predation may explain the random selection of perch site in most birds. Finally, the results indicate that alpine birds may be more favoured by the warming of the alpine region.

Introduction

Thermoregulation, the adjustment of body temperature, is one of the most important metabolic activities in both ectotherms and endotherms (Tansey and Johnson 2015). Maintaining body temperature is fundamental for the majority of metabolic processes as the enzymes that are involved are sensitive to temperature (DiBona 2003). Thermal neutrality is found in endotherms, in which an organism invests the minimal energy to regulate body temperate. Beyond a particular range of ambient temperature, called the thermal neutral zone, organisms have to pay extra costs to adjust; they use behaviours such as evaporative cooling when higher than the upper critical temperature and shivering when lower are commonly found (Zhao et al. 2014). Because of their higher body temperature and basal metabolic rate, birds have relatively higher thermal neutral zones than mammals (Toro-Velasquez et al. 2014).

Thermoregulation can be achieved by physiological and behavioural activities, including vasoconstriction, vasodilation and hormone adjustments. Birds can physiologically maintain their body temperature by adjusting their circulatory system, like the concurrent blood flow in penguin feet on ice and heat exchange from the toucan bill, yet evidence shows that birds also have complex behaviours responding to unfavourable temperatures (Durfee 2008; Tattersall et al. 2009). Several studies suggested that lizards employ basking behaviours for heating and, similarly, birds use radiation from the sun when the ambient temperature is low (Dzialowski and O'Connor 2001). To maintain heat, chickens usually fluff out feathers when standing still, and many small birds huddle in groups; other heat-retaining behaviours such as spreading wings and facing the sun are also found in most avian species (Hafez 1964).

However, there is always a trade-off between thermal benefit and predation risk when perching in the sun, as illumination from the sun is not only linked with heat radiation but also exposes organisms to predators. 'Disability glare' in high light environment may also lead to a delayed response to predators and therefore some birds, like brown-headed cowbirds (*Molothrus ater*), show a preference for shade rather than sunny areas for foraging (Fernández-Juricic et al. 2012). However, because exposure to sunlight significantly decreases metabolic rate, and some birds, such as sunbirds, prefer to forage in the sun (Carr and Lima 2014).

Low temperature and high light are typical characteristics of alpine region, which provides a variety of challenges for local animals. The environment is especially harsh for birds because of their relatively high energetic and thermal requirements, and smaller birds have even higher energy requirements due to their higher surface area to volume ratio (Carrascal et al. 2001). The behavioural thermoregulation of alpine birds is not well known, especially in early summer when most of predators are active but temperature is still low. In this study, we observed and analysed the thermoregulation behaviour of birds in the alpine region, testing the impact of temperature and illumination for perch-site choice among different species. We predicted that birds would prefer warm and high light perching sites, and small birds would prefer a warm and high light site to a greater extent than large birds as smaller birds are quicker to lose heat.

Materials and methods

Study site and species

Our study site was located around the Southern Alps Lodge in Charlotte Pass, New South Wales, Australia. The study site was 300 m in diameter and the elevation was 1,783 m above sea level. The site was a mixture of snow gum forest and grassland, most of which was impacted by human activities.

The bird species we mainly focused on was the little raven (*Corvus mellori*) with an average mass of 428 g. We also observed the behaviour of crimson rosellas (*Platycercus elegans*, 127 g), common starlings (*Sturnus vulgaris*, 80 g) and flame robins (*Petroica phoenicea*, 14 g). The thermal neutral zone of common starlings is 28–37°C (Dmi'el and Tel-Tzur 1985) and 18–32°C for crimson rosellas (McNab and Salisbury 1995). Thermal neutral zones of the little raven (*Corvus mellori*) and flame robin (*P. phoenicea*) are unknown, but the brown-necked raven (*Corvus rificollis*), in the same genus as the little raven, has a neutral zone of 30–37°C (Marder 1973).

Experimental observations were made in early summer, between 14 and 16 December 2015. We went out birdwatching on both the morning and afternoon of 14 December, which was a sunny and relatively warm day. On the following two days, data were only collected in the mornings, and the weather was cold on both days: 15 December was cloudy while 16 December followed a rainy night. We treated 14 December as a warm day and 15 and 16 December as cold days.

Field observation

To explore the conditions of the environment, background readings were made continuously during field observation. Every 3 minutes, two nearby objects were randomly selected as potential perching sites, and temperature and light intensity were measured with a thermal gun (GM320 infrared thermometer) and light meter (Sanwa illuminance meter LX2). The type of object (branch, grass, rock, shrub), temperature (°C), light intensity (Klux) and time were recorded.

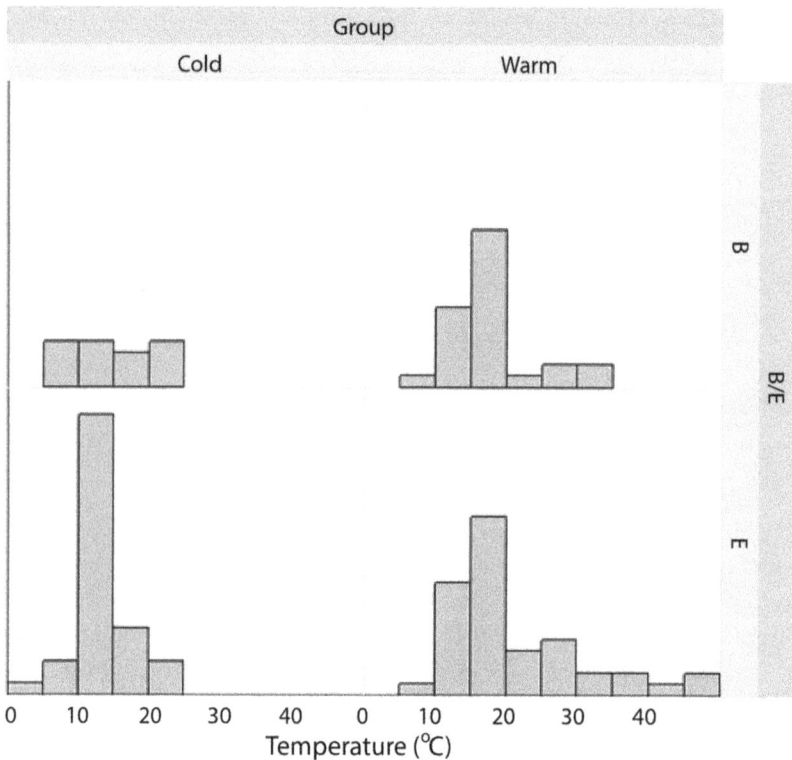

Figure 1: Frequency histogram of temperature comparing perch site and environment on cold and warm days. B represents bird perch site while E represents environment (background reading). Temperature, which was measured by thermal gun, was divided into 5°C intervals (in x-axis) and the height of the column represents the frequency of the object temperature. By comparing the temperature distribution of perch site and environment, it is suggested that birds have no preference based on temperature during warm days but do not randomly select perch site on cold days.

Source: Authors' data.

For birdwatching, those large birds such as ravens, rosellas and starlings that were observed to be sitting for 35 seconds or longer were considered to be perching, while the time threshold for small birds (e.g. robins) was 20 seconds. Temperature of the perching site was measured using a thermal gun or thermal camera (FLIR T420) and light intensity was measured using a light meter. If the perch site was high up in the tree and out of the range of the measuring instrument, the temperature and light immediately beneath the perch site was assumed to be the same and was recorded. Also, time, species, location and the description of bird behaviour (puffing the feathers, sitting with legs hidden, perch in the shade/sun) was recorded.

Data analysis and statistics

Temperature data of thermal images were analysed by FLIR software. The pixels of the perch site were recorded as well as four random spots in the vicinity of the site, which included above, below, left and right of where the bird was. The nearby spots were recorded for microenvironment analysis. All the data were analysed by software JMP10 and a one-way ANOVA test was used for comparing the difference between background and bird perch site.

Results

Overall, we observed 58 birds perching either on branches or on grasses. Ravens comprised 65 per cent of these birds: most were on branches but two were on grasses. Starlings took up 9 per cent of the whole sample and all of them were found on grasses. Rosellas and robins were all found on branches, which took up 14 per cent and 12 per cent respectively. No bird was found on rocks or shrubs and therefore the temperature and light intensity of these substrates was discarded.

When considering the temperature preference of birds, a histogram was made for comparison (Figure 1). The ANOVA showed that, on both the cold and warm days, the mean temperature of the perch site and background were not statistically different. The P value for the warm day was 0.1243 and the similarity was also observed in the shape of the histogram. On the warm day, the temperature distributions of perch site and environment were all positively skewed with a highest frequency of

15–20°C. Although the mean temperature for cold days was not different, the distribution patterns of temperature in perch site and environment were. Despite a normally distributed background temperature, birds on the cold day had almost equal likelihood to occur in all positions on the temperature gradient, which means the selection of perch site temperature was non-random (Figure 1).

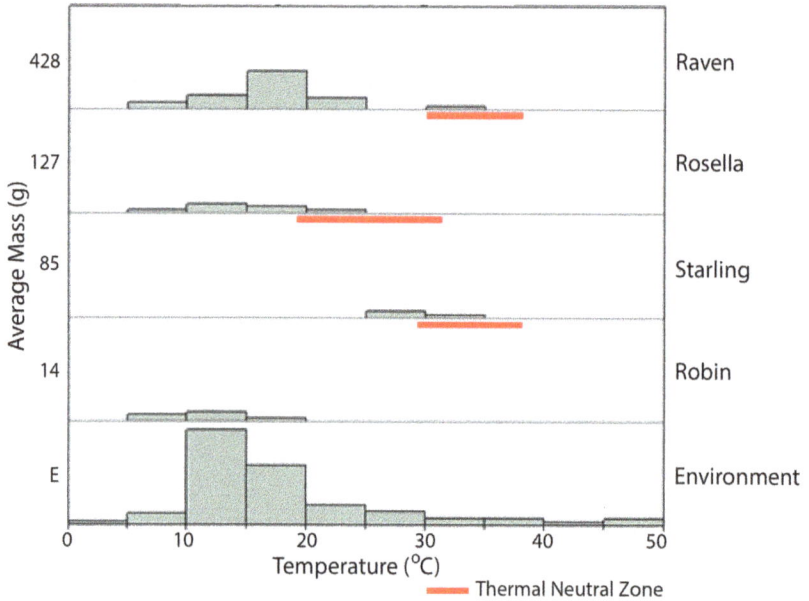

Figure 2: Frequency histogram of temperature by species. Temperature was divided into 5°C intervals (on x-axis) and the height of the column represents the frequency of the object temperature. The average mass of each species is shown on the y-axis. The bottom plot shows the temperature distribution for the environment. The red bars indicate the thermal neutral zone of each species (referenced in the Method section). In total, we observed 42 ravens, 8 rosellas, 3 starlings and 6 robins perching. The three species with red bars show perching behaviour as a response when they are out of their thermal neutral range.
Source: Authors' data.

The preference of temperature also varied among species (Figure 2). The histogram indicated that both robins and rosellas have no preference for temperature of perch site since the distributions are very similar to the distribution of temperatures in the environment. ANOVA confirmed the observation with P values of 0.0916 for robins and 0.2933 for rosellas. Starlings occurred in the place where temperature was higher than background and the ANOVA test is significant (P=0.0232) yet the result needs to be discussed as the habitat of starlings is grassland, which is much warmer than other background objects. Although the ANOVA test for ravens was not statistically significant (P=0.8602), the temperature distribution was different. The perch site temperature of ravens was normally distributed with the highest frequency ranging in 15–20°C while the ambient temperature was positively skewed and most likely to be 10–15°C. Consequently, ravens had a preference for warmer sites to help with thermoregulation. When comparing the frequency with thermal neutral zone (referenced in the Method section), all the birds except robins tended to perch more when the ambient temperature was lower than their lower critical temperature.

Temperature distribution of microenvironment was also analysed. The distributions of perch site temperature were highly similar to the temperature of the surroundings, which was measured with a thermal camera. The ANOVA test also suggested that they were not statistically significant with P values of 0.9443 for cold days and 0.2835 for the warm day. As a result, alpine birds have no preference on particular warmer branches or grasses.

The comparison of light in perch site with environment during the cold and warm days indicated that there was no significant difference with P value of 0.9255 for cold days and 0.0740 for warm day. The P value for the warm day is very close to 0.05 and the ANOVA test indicated a slightly higher light level in perch site. The light preference differed among species (Figure 3). Similar to temperature preference, robins and rosellas had no predilection on light (P value of 0.9591 for robins and 0.6835 for rosellas) while starling only stayed in high-light grassland (P=0.0022, significant). The P value for ravens was 0.3188, which was not significant. However, when comparing to light level of environment, which was mostly below 25 Klux, ravens equally perched in the place that had light intensity below 150 Klux, which suggested that ravens preferred brighter sites.

Behaviours like puffing and sitting were also found when birds were perching. The frequencies of these behaviours as well as the position of perch site were analysed and shown as a bar chart (Figure 4). On cold days, no bird perched in the shade and the frequencies of puffing and sitting were much higher. Smaller birds like robins preferred puffing regulation (50 per cent) on cold days and perching in the sun (60 per cent) on the warm day, whereas larger birds preferred sitting on cold days.

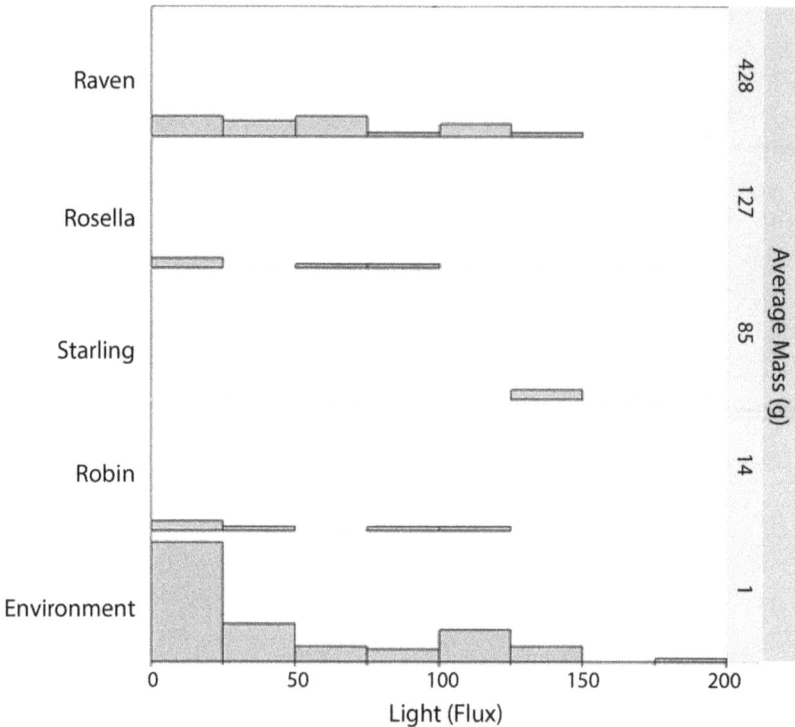

Figure 3: Frequency histogram of light intensity in species. Light intensity was divided into 25 Klux intervals (on x-axis) and the height of the column represents the frequency. The average mass of each species is shown on the y-axis. The bottom histogram (1 g in average mass) represents the light frequency of environment. In total we observed 42 ravens, 8 rosellas, 3 starlings and 6 robins perching. Robins and rosellas show no preference for perch site while starlings perch in high light area. Ravens have a preference for more illuminated sites.

Source: Authors' data.

Discussion

Overall, the results show that only ravens met our prediction of preference for a warm and high light environment. Starlings also significantly chose warm and light perching sites, although the sample size was very small. The hypothesis of particular preference for smaller birds was not supported and the results show that exposure to higher temperatures and light was not linked to body mass. Although there was no particular trend, the selection of temperature on cold days and light on the warm day was not random, and birds did need thermoregulatory behaviours, like puffing and sitting, to help regulate body temperature when they were out of their thermal neutral zone.

The results for the perch site of starlings indicate a statistically significant preference for higher temperature and illumination, yet the difference may not due to thermoregulation. The sample size of starling in our study is very small and limited. All the starlings in the data were found in grasslands that had higher temperatures and light intensity than branches due to sunlight exposure. Other studies confirm that, although considered to be a generalist, starlings in non-urban areas have higher incidence in grasslands than other habitats (Mennechez and Clergeau 2006). Therefore, the high frequency of starlings appearing at high light and warm places is also possibly a result of other factors like nesting and foraging behaviour. Nevertheless, other research also found evidence of seasonal thermoregulation in the red-winged starling, which suggests that starlings in winter have higher basal metabolic rate, resting metabolic rate and body mass to deal with cold environment (Chamane and Downs 2009). The behavioural thermoregulation of the starling is not well known and starlings probably prefer to regulate body temperature physiologically. Future experiments like starling–grassland specific observation are suggested.

Neither robins nor rosellas show preference for warmer and brighter perching sites, which may be due to the risk of predation. The trade-off between thermal benefit and predation risk is always the main factor that impacts thermoregulation behaviour. Several researchers have found that birds appear more responsive to the risk of being preyed upon. Villén-Pérez et al. (2013) suggest that the influence of predation is three times higher than temperature preference for woodland birds in winter and there is no light predilection. Furthermore, some wintering birds even have a preference for shaded areas when dealing with predation risk and change their orientation to avoid glare effect when they are in the sun, which indicates that predation is more important than radiation benefit (Carr and Lima 2014). Therefore, robins and rosellas may consider more about being found by predators and choose not to perch in a warm place with high sunlight. Our study was done in early summer when the predators were more active than during the winter period and so the predation risk was higher, which further reduced the chance of birds perching on warmer and lighter sites. However, for ravens, although large body and black colour make the raven easier to see, there are few potential predators and therefore ravens can utilise radiation without worrying about being eaten.

Another possibility is that the two species, rosella and robin, may prefer to regulate body temperature physiologically rather than behaviourally, yet the physical changes of the birds during cold periods are not well known. Future study of physiological thermoregulation in these small birds is needed. Finally, alpine birds may have acclimated to the cold environment. Solomonov et al. (2009) suggest that birds in a cold environment have a lower metabolic rate and less body temperature stability. Thus the robin, rosella and even raven in the alpine area may have altered their thermal neutral zone to a lower level for cold resistance. The temperature in early summer may not be cold enough for alpine birds to invest in behavioural thermoregulation.

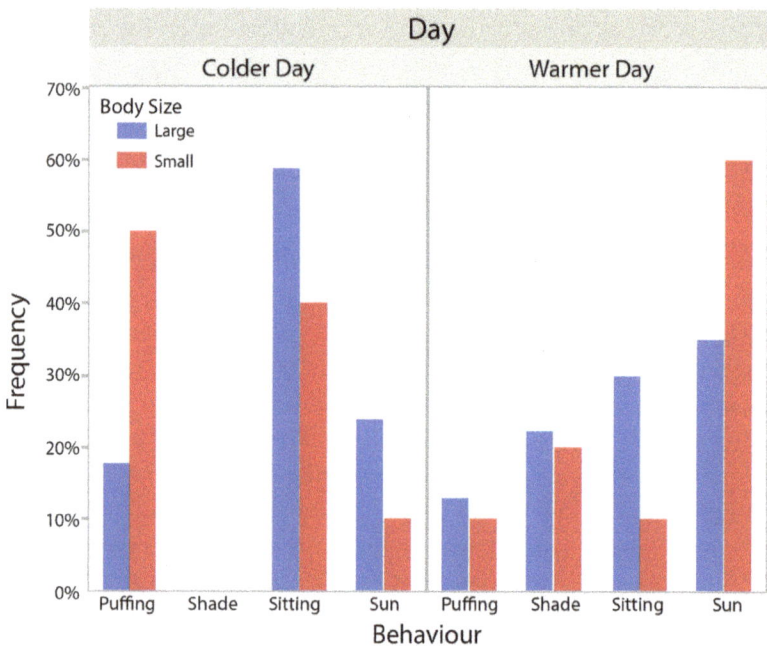

Figure 4: Frequency bar chart of thermoregulation behaviour during warm and cold days. Puffing is the behaviour where birds fluff feathers up and sitting means the bird crouching and hiding their feet. Shade and sun refers to the position where the birds perch. No bird stayed in shade on the cold days and behaviours like puffing and sitting are more likely to be observed on colder days.

Source: Authors' data.

Apart from perching, behaviours like sitting and puffing that we observed are also found in other research. Carr and Lima (2012) identified that wintering birds conserve heat by changing posture like minimising leg exposure and significant ptiloerection, which we recorded as sitting behaviour. These behaviours can be considered as advanced thermoregulation compared to perch-site choice, as sitting and puffing are more active ways to conserve heat. Also, these behaviours do not require solar radiation to gain extra heat, which decreases the risk of being found by predators under the sunlight. Other studies suggest that the surface choice of red-necked nightjars changes due to ambient temperature (Camacho 2013). They are more likely to choose warmer surfaces to land on in cooler weather, which we also found in our study when comparing the temperature preference of birds on warm and cold days.

Future work is highly recommended to refine our results. Firstly, there were measuring errors in this study due to either the measuring instruments themselves or the non-reachable sites. Light meter readings fluctuated due to the change of light intensity when clouds moved. Most of the time birds perched at branches that were high up in the tree and the assumption that the light and temperature of the perch site was same as that of down beneath may be inaccurate. On the top of the tree there was more sunlight and so the temperature and light intensity should be higher than ground, and thermal camera may also have bias as the reading became inaccurate when the objects in the photo was actually far away. Thus, more accurate instruments should be used for further studies. Also, due to the shortage of statistical skills, we used an ANOVA to analyse the differences, which may not be the most appropriate statistical method. The true distributions of temperature and illumination are not consistent with our assumption of normal distribution. And the range of both temperature and light intensity of perch site must be within the range of background data, which makes it hard for an ANOVA test to compare the differences of the modes. Non-parametric methods may be more suitable in this case as the frequencies were not normally distributed. In addition, perching that we observed may not be only due to heat conservation, but also because of feeding and guarding the nest. Thus, longer observation for every bird is suggested for future experiments to eliminate other factors that may cause birds to stay in one place.

This study was done in three days due to time limitation, and a wider range of temperatures is needed to improve the understanding of thermoregulation of alpine birds. We suggest a whole-year observation.

Furthermore, factors like wind speed and orientation of birds can also be measured and analysed so that the perch behaviour can be better explained. The impact of predation can be examined by analysing the potential shelter sites in the surroundings. If the influence of predation risk for alpine birds is more important than radiation benefit, shelter and coverage for them is indispensable, which may indicate that people should avoid destroying the potential shelter for birds when developing the tourist industry. The thermoregulation behaviour we observed indicates that even in early summer the temperature is still unfavourable for alpine birds. Ravens prefer warm perch sites and other birds change postures to conserve heat, which provides us a broad view of thermoregulatory strategy in alpine birds. If the alpine region becomes warmer due to climate change, the birds there will possibly be better able to adapt to the changed environmental conditions.

Acknowledgements

We are grateful to Iliana Medina for supervising, and to Adrienne Nicotra and other BIOL2203 staff for running this wonderful course.

References

Camacho C (2013) Behavioural thermoregulation in man-made habitats: Surface choice and mortality risk in red-necked nightjars. *Bird Study* 60: 124–30. doi.org/10.1080/00063657.2012.753400

Carr J, Lima S (2014) Wintering birds avoid warm sunshine: Predation and the costs of foraging in sunlight. *Oecologia* 174: 713–21. doi.org/10.1007/s00442-013-2804-7

Carr JM, Lima SL (2012) Heat-conserving postures hinder escape: A thermoregulation–predation trade-off in wintering birds. *Behavioural Ecology* 23: 434–41. doi.org/10.1093/beheco/arr208

Carrascal LM, Díaz, JA, Huertas DL, Mozetich I (2001) Behavioural thermoregulation by treecreepers: Trade-off between saving energy and reducing Crypsis. *Ecology* 82: 1642–54. doi.org/10.1890/0012-9658(2001)082[1642:BTBTTO]2.0.CO;2

Chamane SC, Downs CT (2009) Seasonal effects on metabolism and thermoregulation abilities of the red-winged starling (*Onychognathus morio*). *Journal of Thermal Biology* 34: 337–41. doi.org/10.1016/j.jtherbio.2009.06.005

DiBona GF (2003) Thermoregulation. *American Journal of Physiology – Regulatory, Integrative and Comparative Physiology* 284: R277–R279. doi.org/10.1152/ajpregu.00571.2002

Dmi'el R, Tel-Tzur D (1985) Heat balance of two starling species (*Sturnus vulgaris* and *Onychognathus tristrami*) from temperate and desert habitats. *Journal of Comparative Physiology B* 155: 395–402. doi.org/10.1007/BF00687484

Durfee S (2008) Review: Skin: A Natural History by Nina G. Jablonski. *The American Biology Teacher* 70: 56. doi.org/10.2307/30163197

Dzialowski, EM, O'Connor, MP (2001) Physiological control of warming and cooling during simulated shuttling and basking in lizards. *Physiological and Biochemical Zoology: Ecological and Evolutionary Approaches* 74: 679–93. doi.org/10.1086/322929

Fernández-Juricic E, Deisher M, Stark AC, Randolet J (2012) Predator detection is limited in microhabitats with high light intensity: An experiment with brown-headed cowbirds. *Ethology* 118: 341–50. doi.org/10.1111/j.1439-0310.2012.02020.x

Hafez ESE (1964) Behavioural thermoregulation in mammals and birds. *International Journal of Biometeorology* 7: 231–40. doi.org/10.1007/BF02187455

Marder J (1973) Body temperature regulation in the brown-necked raven (*Corvus corax ruficollis*)—II. Thermal changes in the plumage of ravens exposed to solar radiation. *Comparative Biochemistry and Physiology Part A: Physiology* 45: 431–40. doi.org/10.1016/0300-9629(73)90450-7

McNab BK, Salisbury CA (1995) Energetics of New Zealand's temperate parrots. *New Zealand Journal of Zoology* 22: 339–49. doi.org/10.1080/03014223.1995.9518050

Mennechez G, Clergeau P (2006) Effect of urbanisation on habitat generalists: Starlings not so flexible? *Acta Oecologica* 30: 182–91. doi.org/10.1016/j.actao.2006.03.002

Solomonov NG, Anoufriev AI, Isayev AP, Nakhodkin NA, Solomonova TN, Yadrikhinskiy VF, Mordosova NI, Okhlopkov IM (2009) 128. Thermoregulation of cold-adapted birds and mammals of Yakutia (north-east of Siberia). *Cryobiology* 59: 406. doi.org/10.1016/j.cryobiol.2009.10.142

Tansey EA, Johnson CD (2015) Recent advances in thermoregulation. *Advances in Physiology Education* 39: 139–48. doi.org/10.1152/advan.00126.2014

Tattersall GJ, Andrade DV, Abe AS (2009) Heat exchange from the toucan bill reveals a controllable vascular thermal radiator. *Science* 325: 468–70. doi.org/10.1126/science.1175553

Toro-Velasquez PA, Bícego KC, Mortola JP (2014) Chicken hatchlings prefer ambient temperatures lower than their thermoneutral zone. *Comparative Biochemistry and Physiology Part A: Molecular & Integrative Physiology* 176: 13–19. doi.org/10.1016/j.cbpa.2014.06.008

Villén-Pérez S, Carrascal LM, Seoane J (2013) Foraging patch selection in winter: A balance between predation risk and thermoregulation benefit. *PLoS ONE* 8: e68448. doi.org/10.1371/journal.pone.0068448

Zhao Z-J, Chi Q-S, Liu Q-S, Zheng W-H, Liu J-S, Wang D-H (2014) The shift of thermoneutral zone in striped hamster acclimated to different temperatures. *PLoS ONE* 9: e84396. doi.org/10.1371/journal.pone.0084396

Salt-seeking behaviour of mammals and ants in a low-sodium Australian alpine environment

Rachael Robb, Bronte Sinclair, Islay Andrew, Kristi Lee,
Xiaoyun Li, Isabella Robinson, Holly Sargent, Giles Young

Abstract

Salt is an essential nutrient that is found in low concentrations in alpine areas. Salt is thought to be sought by animals through a variety of behaviours including chewing wood, using salt licks, digging and eating plants with high sodium concentrations. We conducted three experiments to investigate whether animals in a low-sodium Australian alpine environment would show a preference for sodium chloride over other salts. Our experiments tested whether 1) alpine mammals would chew on sodium chloride–soaked wooden stakes, 2) ants would be attracted to salt grains and 3) the sodium concentrations in alpine plants are lower compared to plants in other Australian environments. We found that herbivorous mammals, such as rabbits, chewed a greater number of wooden stakes treated with sodium chloride than those treated with other salt solutions. No significant preference was recorded by ants for salt or sugar. Lastly, we found sodium concentration of vegetation decreases with increasing elevations, but the relationship was not statistically significant. Our study was conducted at three locations at elevations of 920 m, 1,200 m, and 1,860 m within Kosciuszko National Park in the Snowy Mountains, New South Wales, Australia.

Introduction

The Snowy Mountains area is reported to have low-sodium soil and vegetation mainly due to nutrient leaching by annual snow melt, moderate precipitation, freezing and thawing (Best et al. 2013). This results from sodium ions in the soil dissolving in water, causing downhill run-off to leach the nutrients from higher elevation soil (Blair-West et al. 1968).

Additionally, the area's low sodium concentration is a result of the lower sodium concentration of rainfall further from the coast (Kaspari et al. 2010). This potentially causes higher elevations to have lower sodium concentrations than at lower elevations. The alpine species in the high elevation regions are suited to their environment with characteristics to cope with the environmental restrictions and conditions of the ecosystem, including the low availability of sodium (Belovsky and Jordan 1981). For example, kangaroos were found to regulate their mineral homeostasis with licks, and in the Snowy Mountains showed physiological adaptations for sodium conservation compared to the Victorian coast (Best et al. 2013). Nutrient cycling in plants is also an example of an ecosystem function that can respond to limited nutrient availability and is thought to be affected by climate change (Díaz and Cabido 1997).

An appetite for salt has been observed globally through selective grazing by sheep, and snow shoe hares gnawing salt-exposed wood (Blair-West et al. 1968). Additionally, salt licks are known to be sought by ruminants, gorillas, butterflies and bees (Kaspari et al. 2008). Supporting studies have suggested that animals seek sodium from vegetation matter when the surrounding soils have been leached of sodium (Belovsky and Jordan 1981), such as mountain gorillas eating decaying wood for 95 per cent of their sodium intake (Rothman et al. 2006). It is not known whether the animal behavioural response to seek salt comes from a need to supplement their diet, a taste for sodium, the low sodium concentration in their environment compared to other environments, or a combination of these interactions (Denton 1982; Belovsky and Jordan 1981; Weir 1972).

Rothman et al. (2006) observed that herbivores with sodium-deficient diets licked or chewed items with high sodium concentrations. Australian herbivores are known to seek high-sodium food while excreting limited sodium (Belovsky and Jordan 1981). Using these observations, wooden stakes soaked in salt solutions have been used to test herbivorous mammals for sodium-seeking behavioural responses (Blair-West et al. 1968). Furthermore, using tubes of sodium solution for baits, ant communities have provided an invertebrate test species and a model system for the biogeographical variation of salt availability (Kaspari et al. 2008; Kaspari et al. 2010). The sodium concentration of vegetation within different environments has been measured in previous studies such as by Denton (1982), who recorded a range of sodium concentrations from 2.25 mmol/kg to 168.75 mmol/kg, with the Snowy Mountains having the lowest sodium concentration. Denton's study was conducted to provide context and determine a reason for animals seeking to increase their sodium intake.

Our study aims to determine the behavioural response of animals to sodium concentration, focusing on the active behaviour of alpine ants and mammals like rabbits and wallabies to sodium and other salts. We also tested the sodium concentration of the surrounding vegetation to determine if and why alpine animals at different elevations actively seek and consume salt in their diet.

The significance of this study is in its application for resource management and conservation (Weir 1972) through revegetation, the control of invasive species like rabbits (Blair-West et al. 1968), and minimising the negative effects of road salting such as the suppression of vegetation growth and germination and increased consumer activity (Kaspari et al. 2010). It will contribute to the knowledge about the Kosciuszko National Park and its inhabitants, and the data collected could be applied to climate change scenarios when considering how animals may respond to temperature increases and shifting seasonal conditions in the low-sodium environment (Díaz and Cabido 1997). It may also have potential connections to studies in other alpine areas where animals may show similar or different behavioural responses to sodium.

We hypothesise that alpine animals will actively seek salt and prefer sodium to other salts. Herbivorous alpine mammals are expected to have greater preference for sodium chloride–soaked wooden stakes rather than stakes treated with other salt solutions, when compared to water-treated stakes. We also hypothesise that invertebrates, such as ants, will show a preference for sodium chloride when compared to white sugar. Lastly, we hypothesise that sodium concentration will decrease with increasing elevation, and alpine plants will be lower in sodium compared to other Australian plants.

Methods

Sites

The experiments were conducted in December 2015 across three sites along an elevation gradient: Waste Point (920 m), Sawpit Creek (1,200 m) and Charlotte Pass (1,860 m). The experiments at Waste Point were conducted near human residences with a lake in the area. Sawpit Creek is a picnic spot near a slope susceptible to road run-off. Charlotte Pass is located in the alpine zone, and the experiments were conducted in a human-occupied resort area near walking paths.

Stakes

Sixty wooden stakes were soaked in salt solutions of either sodium chloride (NaCl), potassium chloride (KCl), magnesium chloride (MgCl), sodium bicarbonate ($NaHCO_3$), or distilled water (H_2O) as a control. One of each treatment formed a five-stake plot, and four plots were set up in each of the three sites: Waste Point, Sawpit Creek and Charlotte Pass. The plots were positioned in areas of likely herbivorous mammal activity, where mostly rabbits and wallabies had previously been observed. Thirty of the 60 salt-treated wooden stakes were placed in the field unmonitored for three weeks, the remaining 30 for two days. Each stake was hammered into the ground for a quarter of its length, to leave a similar surface area exposed. After the test duration elapsed, each site was visited and the stakes in each plot observed for signs of chewing. The chew intensity of salt-soaked wooden stakes was measured using a categorical rating from 1 to 4 indicating none, light, moderate and heavy chewing.

Ants

In the same three sites as the wooden stakes, four Eppendorf tubes were placed around each ant nest, with two active nests per location; two tubes containing 0.25ml of iodised table salt grains, two containing 0.25 ml of white sugar grains. The tubes were arranged in a random order in a semi-circle around the chosen nest, and if needed staked into position with thin wooden skewers. The Eppendorf tubes were left open for 45 minutes then checked for ants inside, which were counted and removed, and then the tube sealed. Before and after placement, the mass (g) of the tube and treatment was measured and the difference calculated. The same procedure was used for 14 control tubes: 10 empty tubes placed at an active nest at the Charlotte Pass site for 45 minutes, and four tubes of salt or sugar with two replicates each placed in an ant-free area within the Charlotte Pass ski lodge. The difference in mass of the Eppendorf tube and treatment of salt or sugar grains by ants was corrected using the measured average mass differences of the control tubes, to allow for the changes in mass unrelated to the ants, such as through added moisture. The correcting difference was subtracted from each tube's original change in mass.

Vegetation

Following the methods used by Denton (1982), we collected foliage samples from each of one dominant grass (*Poa* sp., unidentified), shrub (*Hovea montana*, *Prostanthera cuneata*, *Cassinia* sp., *Cheysocephalum* sp., *Leptorhynchos* sp.), and tree (*Eucalyptus pauciflora*) to provide a mixed representation of alpine vegetation. Enough leaves to constitute a 0.5 g sample of each vegetation type were collected from each of the three sites and placed in labelled paper bags (n=9). The leaves were then dried using a small oven set at 70°C for approximately 12 hours overnight, until the moisture in the leaves was completely removed. We prepared 0.5 g of each foliage sample using a coffee grinder in combination with a mortar and pestle to form a fine powder, and added the sample to a Falcon tube labelled with the plant's scientific name and its collection site. To each of the Falcon tubes, we added 30 mL of distilled water and the tube was shaken vigorously for 30 seconds. Once settled, we removed 30 mL of the extract using a syringe then filtered it into a new Falcon tube with 11 cm filter paper. To get a mmol/kg measurement of the sodium ions in each vegetation sample, we added 3 mL of ionic strength adjuster for 30 mL of extract and then measured using an HI98191 ISE meter with a FC300B Combination Sodium Electrode (Hanna Instruments).

Statistical analysis

To analyse the results of the chewing intensity (categorical rating 1 to 4), removal of salt and sugar (g) and vegetation sodium concentration (mmol/kg), no formal statistical tests were performed because the samples sizes were very small. However, we did calculate the mean, standard deviation and error, confidence interval, degrees of freedom and t-interval of each variable.

Results

At three different sites at different elevations, we tested the intensity of chewing of the wooden stakes, the removal of salt grains by ants and the sodium concentration of leaves. At each of these sites, the salt-treated stakes showed some degree of chewing, ants showed behavioural interest in the salt and sugar grains, and low concentration of sodium was measured in the surrounding foliage.

In the experiment to test herbivorous mammals for salt preference behaviour, wooden stakes soaked with salt solutions showed physical signs of chewing when compared to the control stakes treated with water. The stacked column graph showed that the highest number of stakes chewed were treated with NaCl, which we also observed had the heaviest chewing (Figure 1).

As some of the stakes were in position at the study sites for three weeks and others for two days, we analysed the results separately. The stakes in position for two days did not show visible chewing, which implies that a longer time period may be required for animals to show a behavioural response to sodium. Therefore, we discarded the two-day sample, reducing the sample size to 30 but increasing the precision and accuracy because the stakes included were those that could show a result. We found that 23 of the stakes in position for three weeks appeared untouched. No stakes treated with the control of water were chewed and of the seven stakes that showed signs of chewing, there were three NaCl stakes, two MgCl stakes, one $NaHCO_3$ stake and one KCl stake.

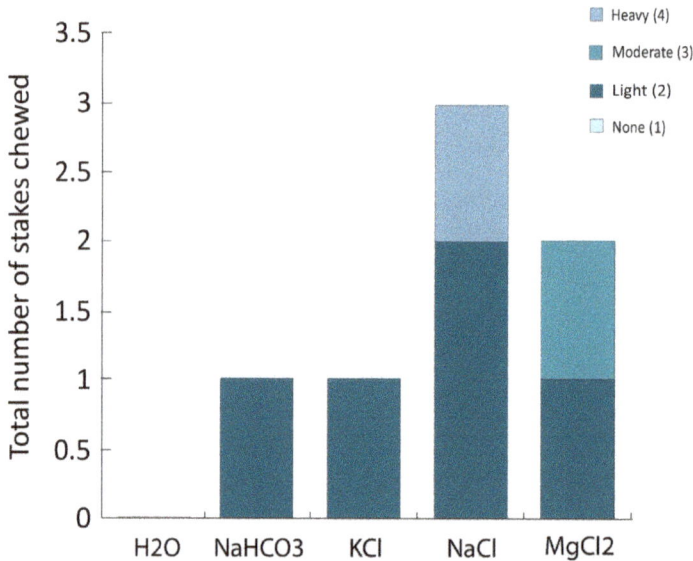

Figure 1: Herbivorous alpine mammals appear to seek NaCl over other treatments. The total number of stakes chewed for each treatment of water (H_2O), sodium bicarbonate ($NaHCO_3$), potassium chloride (KCl), sodium chloride (NaCl) and magnesium chloride (MgCl) was separated into their rating category, with each colour denoting the intensity of chewing.
Source: Authors' data.

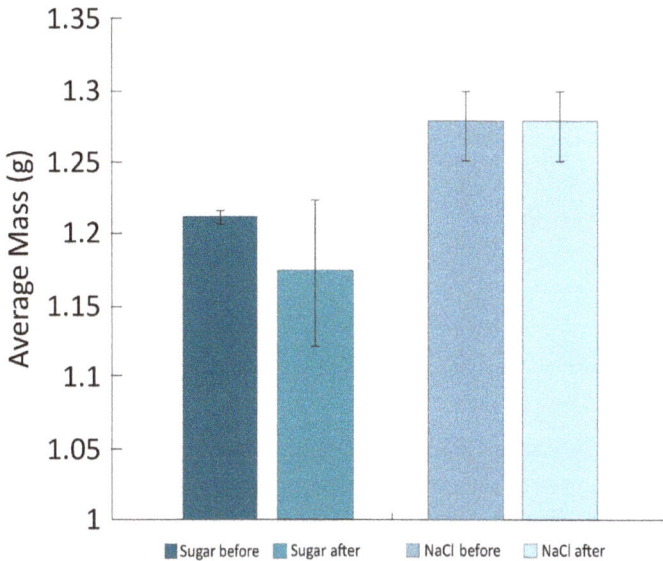

Figure 2: The average mass of the Eppendorf tubes containing table salt (NaCl) or sugar grains before and after placement at the entrance to an active ant nest. Ants did not significantly seek salt or sugar.

Note: Confidence intervals are represented by error bars.

Source: Authors' data.

To see if only mammals showed a preference for salt we tested another animal, using ants as an invertebrate specimen. The average mass of both sugar and salt before and after exposure to ants was not significantly different (Figure 2). Despite this, our observations suggested a preference for sugar when one Eppendorf tube of sugar was completely emptied within the test duration.

At each site, we tested the leaf sodium concentrations of mixed species (n=3, 4) to provide context to the potential reasons why alpine animals might be seeking to increase sodium in their diet. The resulting trend suggests that plants at higher elevations have lower sodium concentration (Figure 3). The error bars show the confidence intervals, indicating no significant difference between the sodium concentrations of the vegetation at the three elevations, despite the observable decreasing trend.

The data taken around Charlotte Pass within the Snowy Mountains was consistent with the data from a previous study by Denton (1982). These areas were shown to have vegetation containing the lowest sodium concentration compared to other areas of Australia (Figure 4).

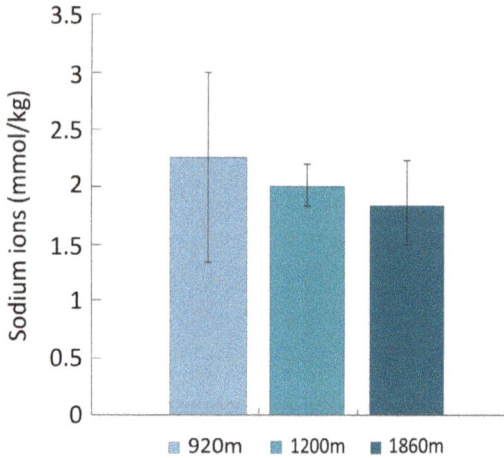

Figure 3: The average sodium ion content measured in a mixed selection of vegetation from Waste Point (920 m), Sawpit Creek (1,200 m), and Charlotte Pass (1,860 m) is shown with error bars displaying the confidence interval. Trend suggests that plants at higher elevations have lower Na+ concentration.

Source: Authors' data.

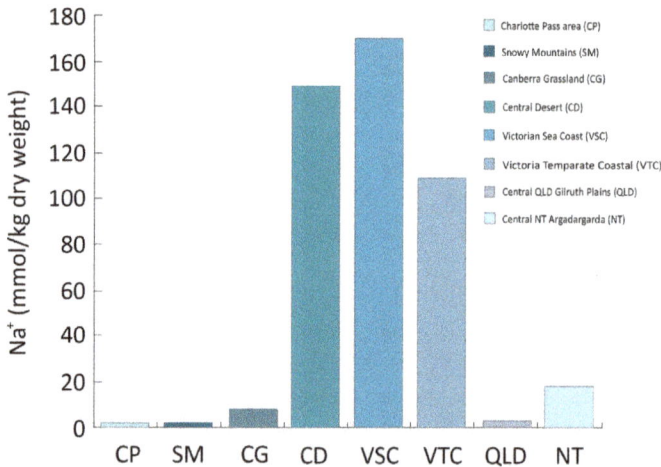

Figure 4: Sodium concentration in Charlotte Pass and the Snowy Mountains is lower than in other Australian regions. The sodium concentration (Na+) data collected in our study of the Charlotte Pass area was compared to the data of a previous study by Denton (1982), which measured the average sodium concentrations in the Snowy Mountains, Canberra Grassland, Central Desert, Victorian Sea Coast, Victoria Temperate Coastal, Central Qld Gilruth Plains, and the Central NT Argadargarda.

Source: Denton (1982) and authors' data.

Discussion

The results showed some preference for sodium by alpine animals and a possible connection to the low-sodium environment; however, this was mostly through trends rather than by statistically significant results. The stakes experiment supported our first hypothesis, but the results from the ant and vegetation experiments could neither confirm nor refute their respective hypotheses.

In the first experiment, the sodium chloride–soaked stakes had the highest number of stakes chewed (3) at the highest intensity of chewing (4) compared to the other salt treatments. This suggests herbivorous mammals in the alpine region have a preference for sodium chloride. Previous observations by Best et al. (2013) similarly recorded a preference for sodium licks over potassium, magnesium or water. Blair-West et al. (1968) also suggested rabbits mostly chewed NaCl stakes over other salt solutions in spring and early summer, the season our study was completed. The highest chewing intensity was at our highest elevation site, coinciding with the theory that higher-elevation alpine environments have lower salt concentrations. This was supported by the third experiment's vegetation having the same range of sodium content to those recorded by Denton (1982), whose measure of Australian plants within the alpine region were much lower than other Australian plants. Other minerals and salts like calcium, potassium and magnesium are possibly attractive to animals as they are also present in the salt licks, soil and vegetation around which animals such as kangaroos and elephants are observed to demonstrate sodium-seeking behavioural responses (Best et al. 2013; Weir 1972).

Sodium can cause other changes in behaviour such as digging, which we observed at Waste Point around the base of the NaCl and to a lesser extent the $NaHCO_3$ stakes, which we suspect was the work of wallabies and rabbits. Digging for sodium is behaviour observed in rabbits, as well as elephants and gorillas (Blair-West et al. 1968; Weir 1972; Rothman et al. 2006). Licking has been another commonly reported behavioural response (Best et al. 2013; Blair-West et al. 1968), but without camera monitoring we could not confirm licking behaviour. Unexpectedly, when we collected the wooden stakes at the end of the two-day duration, we observed a ladybug and a cocoon on the NaCl stake. The observation could not be confirmed as an unconnected occurrence or as one related to the sodium-treated stake.

The stakes treated with magnesium chloride had the second highest number of stakes chewed (2). Blair-West et al. (1968) also found that wild rabbits chewed wooden pegs treated with magnesium, potentially due to magnesium depletion in the rabbits' tissue from sustained high blood aldosterone during summer. The same study observed behavioural interest of rabbits in NaCl and $NaHCO_3$ stakes in spring and early summer, and some chewing of the KCl stakes, which coincides with the light chewing of one KCl stake we observed (Blair-West et al. 1968).

In the ants experiment, a significant change in mass would potentially indicate ants sought to remove salt or sugar from their respective Eppendorf tubes. However, as there was no significant change in average mass, seen by the overlapping confidence intervals, it cannot be confirmed that ants seek salt or sugar. Studies by Kaspari et al. (2008) suggested different ant behaviour than we found in our study; in Kaspari's studied grassland, salt solutions were preferred over sugar by 10–20 per cent of bait usage, and sugar was preferred in coastal areas while further inland the preference was for sodium chloride. These trends contrast with the lack of statistically significant interest we observed ants showing to salt or sugar grains.

The small, black ants tested in our study were thought to be herbivorous, but nothing was known with certainty. As our studied ant species was not identified, the diet and lifestyle of the ants may have been the reason for the different observations. This leads to the style of diet—whether herbivorous or carnivorous—being an alternative but not entirely conflicting theory about why any animal would seek salt, other than a low-sodium environment. Diet is thought to influence sodium bait usage, with herbivores having higher usage than carnivores (Kaspari et al. 2008). This may be due to carnivores being less prone to sodium deficiency because they eat herbivores, which have high tissue sodium content (Blair-West et al. 1968). The concept is not completely conflicting because the diet of herbivores, omnivores and carnivores interact and rely on the availability of nutrients in the environment.

Another possible explanation for conflicting thoughts would be that the mode of presentation of sodium to both the ants and herbivorous mammals may have influenced the results of the experiments. We recommend further investigation into the alternatives to chewing wood and presentation of solid grains, such as isotonic solutions or foodstuff with a high sodium concentration (Blair-West et al. 1968).

The sodium concentrations of alpine plants in our study and Denton's were much lower than published values of other Australian plants, which were as high as 168.8 mmol/kg dry weight on the Victorian coast, but close to the 2.7 mmol/kg average recorded in the central Queensland Gilruth Plains (Denton 1982). The salt content of the plants from the Snowy Mountains given in Denton's study range from 0.1 to 3.8 mmol/kg dry weight, with an average of 2.25 mmol/kg. The sodium values we measured, an average of 2.04 mmol/kg, were within this range.

Human activity has influenced sodium supply and animal distribution (Jensen et al. 2014; Kaspari et al. 2008). Road salting increases the natural salt levels in the environment and chloride levels increase the pH of the soil, exacerbating the already acidic soils at high elevations (Jensen et al. 2014). The increase in salt in places where road run-off is high can harm native vegetation growth and nutrient uptake, suppressing seed germination as well as polluting local waterways with chloride ions (Kaspari et al. 2010). Higher salt concentration along roads would also promote consumers (Kaspari et al. 2010), potentially increasing the risk of roadkill by drawing animals closer to the road in the search for food with higher sodium content. Invasive animals would be influenced by road salting; through scat observations, we thought rabbits were the animal most commonly chewing on the stakes. Increasing the salt concentration through salting could influence rabbit population density and distribution, potentially leading to concentrated damage to soil quality and native vegetation communities as herbivory is increased (Forsyth et al. 2015; Kaspari et al. 2010). A similar occurrence was reported in African elephants, whose distribution and increase in population density was concentrated where the water and soil have high sodium content (Weir 1972).

Errors and improvements

The method of rating the chew intensity of each stake into categories was susceptible to bias and inaccuracy, leading to random errors. This could be minimised by using a numerical method of measuring the chewing, such as scanning the surface area or weighing each stake before and after the sample duration (Blair-West et al. 1968). This would be further improved and supported by introducing camera monitoring for each plot, such as used in the study conducted by Best et al. (2013), to more accurately observe and record which animals chewed the stakes. This information would allow more accurate categorisation of the activity at each stake, including the duration of visit, how many animals visited and identifying other indications of the animal salt-seeking activity, such as licking and digging. Monitoring would help address the unknowns in our study, such as which animals were responsible for the marks on the stakes and the digging found at the base of some stakes. The type of wood used for the stakes could also be better controlled if further investigation was done; perhaps consistently using softer wood to better show chew marks (Blair-West et al. 1968).

The method of testing for sodium or sugar preference using the average mass differences was flawed. This was because the difference caused by the removal of one or two grains was not significant, the method of presentation and concentration of salt possibly did not fit with an ant's natural diet, and the mass would be unbalanced by interfering factors like moisture or dirt collecting inside the tubes. Therefore it is likely that the results would not show a preference if there was only a slight difference between the initial and end mass. It also did not allow for behaviour like the ants emptying a tube containing the sugar treatment to contribute significantly to the data. The time factor of 45 minutes limited the ant experiment as the ants may not have been continuously active; natural diet could dictate the 0.25 mL of salt would not be needed at once, and so they removed the small amount that satisfied their requirement and then left. We also observed that there was less activity at collection than when the tubes were set up. Considering these factors, if tested over a period of days or weeks, it would be likely that more salt and sugar would be removed overall as the ants may return to the known nutrient sources.

Increasing the replicates would improve all three experiments; through increasing the duration of each experiment, having an elevation gradient with more than three sites and a larger variety of study animals, the results would more likely be statistically significant. Additionally, we could increase the number of stake plots and nest samples, collect a greater range of vegetation samples, as well as test the soil and other plant components for sodium.

Another contribution to further study would be to run the ant experiment with other known species of ants to see their reaction to the salt and sugar treatments. This would allow us to collect more information, including what the ants' natural diet included or lacked (Kaspari et al. 2008).

Applications

Our study of alpine animals' responses to sodium in the environment has applications to the management of revegetation, as the choice of flora for alpine environment restoration would need to possess the physiological and functional traits that contribute towards resilience to low-sodium soil. The flora would also be required to survive the other environmental and climate conditions of the region such as seasonal availability of nutrients, snow duration and tolerance to temperature extremes, soil chemistry and moisture availability (Blair-West et al. 1968; McGill et al. 2006).

Significance

The combination of stakes, ants and vegetation testing related to animal behavioural responses to sodium has not been previously completed in the elevation gradient from Waste Point to Charlotte Pass. Therefore, our study provides new information about the area and data that can be compared to existing studies of similar or different environments. This contributes towards a greater understanding of the patterns of sodium distribution and animal response and its application in practical management and conservation.

The low sodium concentration we measured in the vegetation does suggest that animals seek sodium because of a deficiency in their environment. However, for a definite conclusion of whether alpine animals seek salt, and for what reason, more research would have to be done into other animals' behaviour and diets (Blair-West et al. 1968). Further clarification would also be needed of the low sodium concentration in the alpine environment compared to other ecological, climatic and seasonal conditions that could influence animals seeking salt (Weir 1972).

Knowledge of salt and nutrient availability in the alpine environment has management and conservation applications. Therefore, as our study has helped to increase understanding, it has potentially increased the successfulness of alpine management.

Acknowledgements

We would like to acknowledge and thank Hannah Windley and Adrienne Nicotra for their resource assistance and support.

References

Belovsky GE, Jordan PA (1981) Sodium dynamics and adaptations of a moose population. *Journal of Mammalogy* 62: 613–21. doi. org/10.2307/1380408

Best EC, Joseph J, Goldizen AW (2013) Facultative geophagy at natural licks in an Australian marsupial. *Journal of Mammalogy* 94:, 1237–47. doi.org/10.1644/13-MAMM-A-054.1

Blair-West JR, Coghlan JP, Denton DA, Nelson JF, Orchard E, Scoggins BA, Wright RD (1968) Physiological, morphological and behavioural adaptation to a sodium deficient environment by wild native Australian and introduced species of animals. *Nature* 217: 922–8. doi. org/10.1038/217922a0

Denton DA (1982) *The Hunger for Salt: An Anthropological, Physiological and Medical Analysis.* Springer-Verlag, New York.

Díaz S, Cabido M (1997) Plant functional types and ecosystem function in relation to global change. *Journal of Vegetation Science* 8: 463–74. doi.org/10.2307/3237198

Forsyth DM, Scroggie MP, Arthur AD, Lindeman M, Ramsey DSL, McPhee SR, Bloomfield T, Stuart IG (2015) Density-dependent effects of a widespread invasive herbivore on tree survival and biomass during reforestation. *Ecosphere* 6(4): 1–17. doi.org/10.1890/ES14-00453.1

Jensen TC, Meland S, Schartau AK, Walseng B (2014) Does road salting confound the recovery of the microcrustacean community in an acidified lake? *Science of the Total Environment* 478: 36–47. doi. org/10.1016/j.scitotenv.2014.01.076

Kaspari M, Chang C, Weaver J (2010) Salted roads and sodium limitation in a northern forest ant community. *Ecological Entomology* 35: 543–8. doi.org/10.1111/j.1365-2311.2010.01209.x

Kaspari M, Yanoviakc SP, Dudley R (2008) On the biogeography of salt limitation: A study of ant communities. *Proceedings of the National Academy of Sciences of the United States of America* 105: 17848–51. doi. org/10.1073/pnas.0804528105

McGill BJ, Enquist BJ, Weiher E, Westoby M (2006) Rebuilding community ecology from functional traits. *Trends in Ecology and Evolution* 21: 178–85. doi.org/10.1016/j.tree.2006.02.002

Rothman, JM, Van Soest PJ, Pell AN (2006) Decaying wood is a sodium source for mountain gorillas. *Biology Letters* 2: 321–4. doi.org/10.1098/rsbl.2006.0480

Weir JS (1972) Spatial distribution of elephants in an African national park in relation to environmental sodium. *Oikos* 23: 1–13. doi.org/10.2307/3543921

A rose in any other shade: Is alpine flower pollinator distribution driven by colour?

Hannah Zurcher, Ming-Dao Chia, Julia Hammer

Abstract

While some plant–pollinator relations are born of scarcity, competition or necessity, the stunning diversity of Australian alpine flora lends itself to rich fields of varied blooms. If there is any plant–pollinator selection happening in the Australian Alps, then it is likely due to insect colour preference. Insects often prefer flowers of specific colours, as visual cues provide guidance for resource distribution. Bees generally prefer yellow flowers, flies white flowers and beetles white flowers.

There are more white flowers (53.5 per cent), than yellow (21.3 per cent) in Kosciuszko National Park. *Oxylobium ellipticum*, the common shaggy-pea, is a yellow flower found near Charlotte Pass. *Epacris paludosa*, a star-shaped flower, is white and found in a range of alpine areas. *Olearia phlogopappa*, the dusty daisy-bush, blooms both purple and white and is found over much of Australia.

Alpine insects are more restricted than temperate insects in terms of food sources, due to year-round lack of both animal and plant food sources. In spring, then, when flowers bloom, insects take every opportunity they can to feed. Flies are the most common pollinator in Kosciuszko National Park. Bees are more specialist feeders than beetles and flies. Thus, bees generally prefer flowers with more closed structures. Beetles and flies generally prefer more open-structured flowers. Insect pollinators generally prefer flowers with stronger scent signals, and honey bees have been shown almost never to land on a flower without both a colour and scent stimulus.

Given that floral feeding is an important part of the life cycles of alpine insects, pollination selection will differ widely between insect species. We hypothesised that pollinator visitation will differ significantly in terms of morphotype and number of pollinators between Australian alpine native species *O. ellipticum*, *E. paludosa* and *O. phlogopappa*.

Despite visible trends towards pollinator selection, no significant difference could be determined overall. Greater time, more varied conditions, surveys of present insect species and control for plant height (e.g. beetles fly less than bees, so may pollinate yellow plants low to the ground but not higher yellow plants, leading to misleading results) could all refine discussion of alpine plant–pollinator relationships.

Method

We located a field site in subalpine woodland that contained patches of *Oxylobium ellipticum*, *Epacris paludosa*, *Olearia phlogopappa*, with at least one of each in flower and of a structural density that did not leave the plants obviously abnormally sparse.

We observed all three flower species for at least 10 minutes respectively as a group to identify common morphotypes of pollinators, agreed on a system of labelling and generally assured that variations in observational techniques between individuals were limited.

We collected relevant field data: average wind speed and current temperature using a Kestral at both the beginning and end of each half-hour measuring period.

Over a 30-minute period, observers recorded all flower visitations and feedings to the patch of one of the species studied. Each species of plant was observed by at least one group member between each observing period. Morphotype of pollinator was recorded each time it landed on a flower. Whether it fed on the flower or simply visited was recorded.

Each subsequent half-hour observation period was done in a separate location to any of the previous. Researchers attempted to observe a reasonably equal number of the three different plant species over all of the different observational periods.

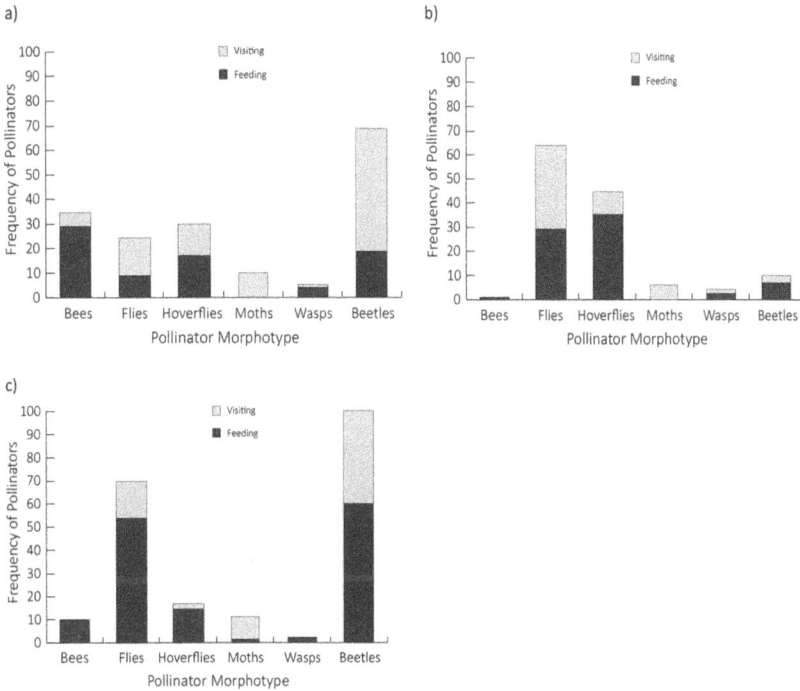

Figure 1: Frequency of pollinator morphotypes on a) *Oxylobium ellipticum*, b) *Olearia phlogopappa*, and c) *Epacris paludosa*. Dark grey bars represent feeding frequency, and light grey bars represent visiting frequency; for example, on *O. ellipticum* 20 beetles fed and 49 beetles visited.

Source: Authors' data.

Results

Figure 1 compares the frequency of each pollinator morphotype (defined as bees, flies, hoverflies, moths, wasps and beetles) on *O. ellipticum*, *O. phlogopappa* and *E. paludosa*. Landing on flowers (visiting) was distinguished from feeding. Figure 2 emphasises pollinator preference for the three flowering plant species.

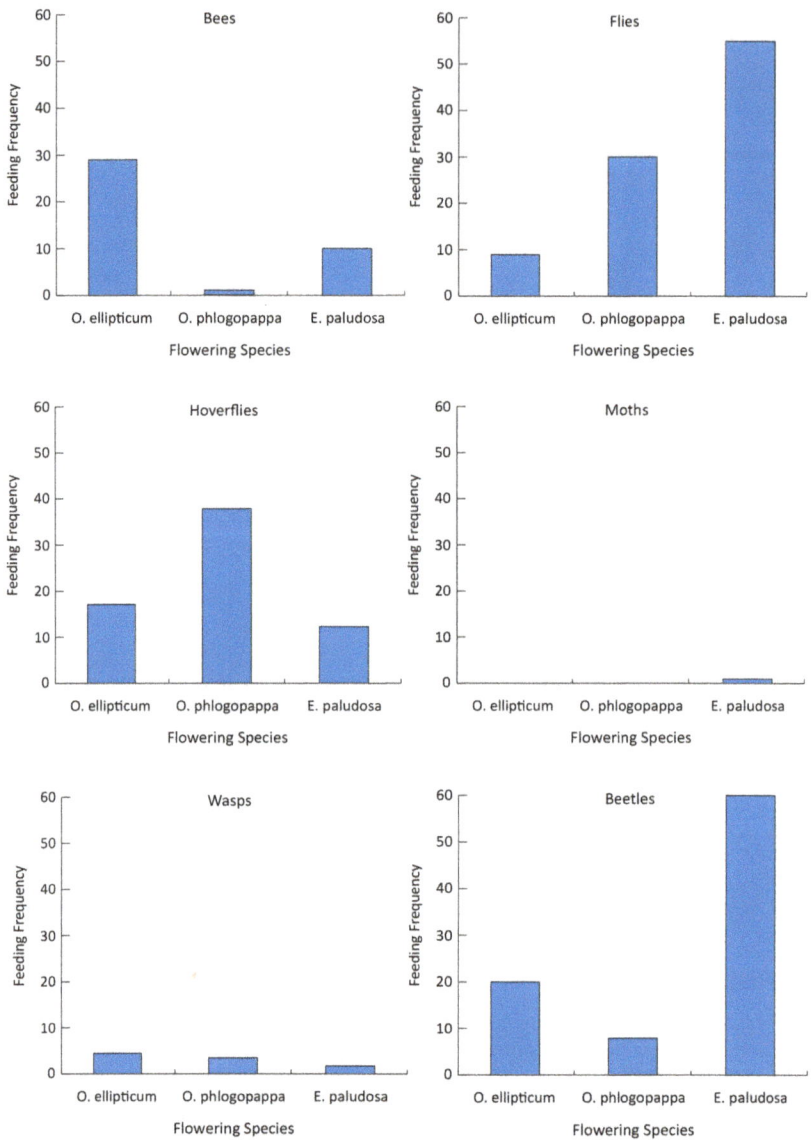

Figure 2: Frequency of feeding by insects for the three focal plant species.
Source: Authors' data.

Discussion

This project was conducted for just two days and then altered because of weather. The notes here are provided as a summary of what was done.

Despite visible trends towards pollinator selection, no significant difference could be determined overall. Greater time, more varied conditions, surveys of present insect species and control for plant height (e.g. beetles fly less than bees, so may pollinate yellow plants low to the ground but not higher yellow plants, leading to misleading results) could all refine discussion of alpine plant–pollinator relationships.

Acknowledgements

We thank Micheal Whitehead for his support and advice, and for supporting our decision when the weather changed and we decided to pull the plug on this project.

References

Costin AB, Gray M, Totterdell CJ, Wimbush DJ (2000) *Kosciuszko Alpine Flora*, 2nd edn. CSIRO Publishing, Melbourne.

Dafni A, Potts SG (2004) The role of flower inclination, depth, and height in the preferences of a pollinating beetle (Coleoptera: Glaphyridae). *Journal of Insect Behaviour* 17: 823–34. doi.org/10.1023/B:JOIR.0000048991.45453.73

Gollan JR, Ashcroft MB, Batley M (2011) Comparison of yellow and white pan traps in surveys of bee fauna in New South Wales, Australia (Hymenoptera: Apoidea: Anthophila). *Australian Journal of Entomology* 50: 174–8. doi.org/10.1111/j.1440-6055.2010.00797.x

Inouye DW, Pyke GH (1988) Pollination biology in the Snowy Mountains of Australia: Comparisons with montane Colorado, USA. *Australian Journal of Ecology* 13: 191–210. doi.org/10.1111/j.1442-9993.1988.tb00968.x

Lord, JM (2008) A test for phylogenetic conservation in plant-pollinator relationships in Australian and New Zealand alpine flora. *New Zealand Journal of Botany* 46: 367–72. doi.org/10.1080/00288250809509774

Pickering CM, Stock M (2004) Insect colour preference compared to flower colours in the Australian Alps, *Nordic Journal of Botany* 23: 217–23. doi.org/10.1111/j.1756-1051.2003.tb00384.x

Research-based learning: Designing the course behind the research

Elizabeth A. Beckmann, Xénia Weber, Michael Whitehead, Adrienne Nicotra

It is hoped that the papers in this book have encouraged readers to share the authors' appreciation of the ecology of Kosciuszko National Park— its diversity, variability, subtlety and uniqueness. The authors also hope readers have been intrigued, challenged and informed by the research problems tackled in the papers, as well as the scientific responses and solutions to those problems. This chapter explains a little more of the way in which this research came about, and makes some connections across some key findings. It also explains the pride that the course design team (the authors of this chapter) have in the work of the papers' authors, most of whom had just a couple of semesters of university study when they embarked on this research.

A vision for research-based learning

In mid-2015, Adrienne Nicotra, Elizabeth (Beth) Beckmann, Xénia Weber and Michael Whitehead came together to plan how to realise Adrienne's vision of bringing together a group of Australian National University (ANU) science students, who had just finished their first year of study, to help them share their enthusiasm, knowledge, skills and energy in a research-based field ecology course. This teaching design team was diverse: Adrienne, a plant science researcher and renowned educator with extensive knowledge in her field and beyond; Beth, a biologist and environmental communication specialist turned educational designer and researcher; Michael, a recent PhD graduate in evolutionary ecology with a passion for teaching; and Xénia, an Honours year science undergraduate experienced in running peer-assisted learning sessions for first-year students. As the team shared ideas and experiences in the context of Adrienne's vision, the structural and pedagogical elements of the course became clear. The plan was for a two-week residential course in Australia's Snowy Mountains, specifically within the high-elevation country of

Kosciuszko National Park.[1] This was finalised into a fortnight in early December (after the end of the academic year, and at the beginning of summer) at Charlotte Pass, just four hours from the university campus. The logistics of travel, residential living and cost in this mildly isolated spot would be complex but manageable.

In terms of its approach to learning, the course was designed to be constructivist (Honebein 1996) in its intention of creating an intellectually challenging research-based educational environment (Healey 2005). It was integral to the pedagogical design that the students would be the primary researchers: identifying research problems inspired by the ecological contexts around them, designing experimental approaches, collecting data, analysing that data in the context of the research problem and presenting their findings, both orally and in writing, to their peers. The proposed course was innovative and definitely ambitious. In an approach that constitutes the research-based education equivalent of rapid prototyping, small teams of students would work through four different research projects over 12 days, iteratively practising and developing their experimental and critical-thinking skills again and again in different contexts.

Crucially, the field expertise of experienced researchers was an important resource to be shared with the students. These 'resource people' were instructed to support and gently guide the students, rather than provide prescriptive leadership based on their expertise. This model drew on Beth's understanding of authentic learning in field courses (Beavis and Beckmann 2012), and especially on Adrienne's experience of a field course run by the Organisation for Tropical Studies during her own doctoral studies.

---

1   Kosciuszko National Park (KNP) was named by the Polish explorer and 'discoverer' Strzelecki in honour of his national hero General Kosciuszko. At 2,030 m, it is the highest mountain in Australia. Within KNP, at 1,760 m above sea level, Charlotte Pass is the highest permanent settlement in Australia and provides relatively easy access to Mt Kosciuszko. It is crucial to remember, however, that KNP and all the high-elevation environments were well known to the local Indigenous people for thousands, if not tens of thousands, of years before such 'discoveries'. The authors are heartened that in 2016 the NSW Government signed a Memorandum of Understanding with the Monaro Ngarigo people, solidifying the local Indigenous community's role in preserving KNP's cultural as well as natural heritage. The authors acknowledge the local Indigenous people, and especially pay respect to their elders past, present and emerging.

Although the team initially had some concern as to whether ANU staff researchers would be convinced by the model and able to 'let go' when working with early year undergraduates rather than doctoral researchers, these concerns proved unfounded. A few days in the alpine environment turned out to be a great drawcard for ecology-minded colleagues at ANU, with senior PhD students and postdocs joining research fellows in volunteering their expertise and enthusiasm. All appeared to find it very natural to focus on the undergraduates driving each project, as they identified their own developing passions and motivations in ecology.

As the design team realised that the course was developing along quite innovative lines, the decision was made to seek university human research ethics approval (ANU Protocol 2015/553) to allow the course to be framed within its own scholarly pedagogical research context and the outcomes more formally assessed and disseminated. (This is planned for future journal papers, as well as through this overview.) The team focused especially on developing thoughtful student feedback surveys, which included creating comparable data sources by adapting questions from published research (Durrant and Hartman 2014; Howitt et al. 2014). A mid-course whole group discussion, facilitated by Beth as the non-residential 'outsider', gave students an option to provide formative feedback during the two-week stay, which led to useful changes for the second half of the course, as well as complementing the other sources of feedback to guide the design of future iterations of the course.

Crafting the workshops

With some key structural elements in place, the next step was to ensure that creativity of the approach was grounded in evidence-based pedagogical strategies. Given the time constraints of an intensive course, students would need to hit the ground running. Some pre-residential readings were considered essential to ensure a reasonable common background in relevant ecological and methodological topics. Beyond theoretical academic knowledge, the team wanted the students to consolidate other crucial scientific skills, such as teamwork, collaboration and communication. To do this, 'just in time' workshops were planned for delivery during the field trip. Xénia's background in peer-assisted learning came to the fore here, and she and Michael began to develop some of the innovative workshops that would take place on-site. Modelling what was intended to happen in the field, Adrienne and Beth became the educational 'resource people' and mentors for Xénia and Michael, encouraging a shared reflective practice and stepping into the background as these two identified and addressed their own research areas. Xénia's reflective journal shares her thinking at the time:

> Group work and reflective practice were two important skills I was particularly keen on fostering among the students. I was also convinced of the importance in tackling these skills explicitly rather than implicitly, because I wanted us to take a different—better—approach than how I'd been taught. My experiences in terms of 'reflection' at school had been largely negative: there was no real explanation of the value of reflection or what good reflective writing actually looked like. Similarly, my experiences associated with group work had been poor. Either staff encouraged (or mandated) group activities and hoped that positive student outcomes would emerge from the opportunity, or teachers would simply recite key qualities associated with good group work (such as letting everyone participate, or respecting differences of opinion in a constructive way). There was never any in-depth exploration of how or why particular behaviours emerge in teams, or what strategies could be used to deal with these behaviours.

> As we discussed options for the ecology course, I realised we could offer unique workshops in reflection and group work that would offer students new material, grounded in current research. This latter point became a real focus, especially after a chance discussion with Andrew Frain[2] in the ANU Research School of Psychology, who described how he was using social identity concepts to workshop collaboration and leadership skills. We were working with two overlapping identity groups—'university students' and 'ecological researchers'—so it made sense to present them with the current developments and insights of psychological researchers into group dynamics, individual perceptions of identity and motivation, including the elements of complexity and uncertainty.

As thoughtful educators, Xénia and Michael took on the task of steeping themselves in the relevant knowledge before designing the workshops. First, they developed a literature review and research database, and learned about others' experiences with implementing such projects in tertiary education. From this background, they distilled the kinds of workshop that they wanted to create, and we all brainstormed how we could connect

---

2   Dr Andrew Frain is now Senior Evaluation Analyst, Planning and Performance Measurement and Teaching Fellow in the Strategic Defence Studies Centre at The Australian National University.

these to other elements of the course (e.g. through a reflective journal and post-workshop surveys) to enhance learning outcomes and evaluate the teaching. Again, Xénia's reflections provide a powerful insight into this planning phase.

> In preparing to teach about collaboration, Michael and I met regularly with Andrew [Frain] in a form of tuition/book club where we would discuss Haslam's social identity approach (2004) and Andrew's own blog.[3] These opportunities were invaluable for posing questions, challenging and eventually understanding the theories on social identity, group/individual motivations and how these could be best presented to students to develop their collaborative skills. This dramatically altered our initial plans of focusing on topics such as 'working in teams', 'active listening', 'conflict resolution' and 'appreciating and benefiting from diversity'. Instead, we sought to understand ways in which we could convey to students how individual/group identities form, how this affects attitudes/behaviour, the implications for functional/dysfunctional groups and opportunities for improving group motivation and collaboration by understanding the underlying processes.
>
> Similarly, although I had used reflective writing many times throughout my academic studies and teaching roles, I sought a stronger basis for how best to instruct and make the process of reflection more accessible and transparent to those who might be less familiar with, or confused about, the genre. I read numerous papers and resources, mostly recommended by Beth, and found the work by Moon (1999) and by the ANU team of Howitt et al. (2014) particularly helpful in providing practical strategies to teach high-quality reflective practice.
>
> As we began to develop the workshop materials, we realised that repeated exposure and interaction with the theories of reflection and collaboration would be critical for our students to understand and 'own' all these concepts. We therefore created a one-page handout to highlight to prospective students how our approach to collaboration would be different, the basis of the social identity approach and why we thought it would be so influential in the course. We also explained these concepts when we introduced the students to their field notebook assessment task, outlining our expectations and the emphasis on reflective quality (rather than quantity), with guiding questions to help reflection.

---

3   Tame, R, Frain, A (2015) *Social identity resources*. socialidentityresources.com/.

The final outcome of this development approach was a set of three workshops[4] that introduced and then built on concepts of self-categorisation (as individuals and then as groups), social identity, effective collaboration (and how this could be supported or antagonised) and reflective writing (as a practical skill for recording a researcher's activities and as a tool to enhance critical thinking). There was much thought put into presenting these concepts, appropriately contextualised to provoke humour and engagement. For example, recognising the strength of imagery in effective communication but unimpressed with the typical business focus of social identity concepts in published research, Xénia and Michael instead came up with the traits of the 'busy bird' and the 'lazy sloth' as a discussion point around group dynamics. Despite the apparent simplicity of this analogy, which initially concerned some students, these terms rapidly became a shorthand readily used by students in their written reflections. Indeed, some were even overheard referring to themselves or their team members going into 'sloth' or 'bird' mode during the course of the day's work. This approach helped students gain a much more personalised understanding, evidenced in their individual reflective journals, of the waxing and waning of an individual's roles and contributions in a team that must work intensely and consistently together on a single project over several days.

The research outcomes

By all indicators, the course was a great success. The students rose beautifully to the task of leading, designing and communicating their research, and improved their skills and elegance across all these aspects as they moved through the four projects. The availability of high-quality digital cameras (mostly in their mobile phones) was put to great use to record data. The quality of their research outcomes can be seen throughout this book. While the work of not yet polished researchers, and naturally limited in scope, the formal peer-review process ensured that all these papers report reliable findings that may have otherwise remained unknown, and that have the potential to guide future researchers towards interesting problems.

---

4   See Workshops 3, 4 and 7 in the 'Workshop summaries' section.

Science holds 'firsts' in high regard. This volume reports the first study on circadian rhythms in the iconic snow gum (*Eucalyptus pauciflora*), where Mauger et al. show convincing evidence for diurnal rhythms in photosynthesis as measured by stomatal activity. In another first for an iconic alpine species, Zucher et al. test the hypothesis that tiling aggregation in bogong moths is a behavioural adaptation to mitigate water loss. While the experimental results run opposite to this compelling hypothesis, this only deepens the mystery surrounding this curious behaviour.

In a changing climate, Australia's alpine ecosystems are likely to be some of the most vulnerable. This concern is addressed by student teams in several of the studies reported in this volume. McLeod et al. contributed evidence of local acclimation of photosynthesis in snow gums. In their studies on alpine skinks, and perhaps controversially, both Hammer et al. and Robinson et al. suggest that 'some like it hot', and that these reptiles might actually benefit under projected temperature increases. Local data and results like these are crucial for refining climate change models.

The course's reception by students

The three workshops on identity, collaboration and reflective practice were particularly successful in providing the students with practical skills and new perspectives on thinking about themselves and their own collaborative interactions in terms of human social behaviour. Again, it is instructive to draw on the reflections of Xénia as the educational researcher that she had become:

> Short presentations were crucial after a long day in the field. An hour was ample to cover all material, and the evening timeslots made the small group interactions and discussion more relaxed and helpful in terms of planning the next day. While behavioural and conceptual 'worldview' learning outcomes are particularly challenging to measure, I was happy with the material we delivered and felt it was a unique and important contribution to the students' learning. To be truthful, I had expected some critique and apprehension about these 'different' ideas, but instead the post-workshop anonymous feedback was exceptionally encouraging. With just the few hours of the workshops, and the encouragement to continue reflecting, many insights seemed to be emerging among the students in regard to their working relationships and their observations in the field. It became clear that the concepts being presented were actually influencing students' ideas and behaviour. Even over the first week, I saw evidence in the students of increasing self-awareness, analytical skills

and an expanded repertoire of potential explanations to social dynamics. I myself am pleased to have learnt and thought about these concepts, and can see the value in pursuing them more myself, especially integrated into other teaching roles.

Thinking and writing reflectively on nature is a well-known characteristic of ecologists—influential writers such as Aldo Leopold, Tracey Storer, Konrad Lorenz and Rachel Carson spring to mind—so it was in keeping with the research-led focus to build on the reflective practice aspects in the assessment schedule. As well as individual field notebooks that required daily entries of notes and data, students were asked to write at least three reflective contributions in their journals each week. With many students writing even more frequently than requested, these reflective journals provided an amazing wealth of written and visual commentary on the individual learning pathways, research discoveries and lessons from residential and collaborative relationships and experiences that the course had facilitated. That the students had especially learned to value reflective practice as learners and scientists was evident in the anonymous post-course feedback, as the below examples show:

> I think reflection and also keeping a field diary is incredibly helpful because you can look back and see what you've learnt or what needs fixing, or how certain data can be helpful even if it's not for what you first thought you were studying or had envisaged your results to say.

***

> … I still have a lot to learn when it comes to reflective practice. However, over the course of just 2 weeks I have gone from being a complete skeptic to appreciating the value of reflection, both personally and scientifically.

***

> The focused workshops and field journal practices on reflection did help in the research and investigation and analysis processes. Having an immediate way to apply that reflection—to the field problems and our field books—gave relevance to our study of reflective practice and helped us to see the significance of it, and solidified the skills, more so than if they were simply mentioned [in] a lecture and never followed up with application.

***

> It was good to have a sense of how I was coping with the course through writing down my experiences …

Collecting such a range of student feedback—formal from post-course surveys and informal from on-site conversations, the mid-course focus group and students' own reflective journals—was very valuable and confirmed the design team's sense of success. Interestingly, the main issue from the mid-course group discussion was the way the evening cooking/cleanup groups were structured, rather than anything to do with the research aspects, reminding everyone of the importance of having a holistic view when designing and implementing residential courses.

In the post-course surveys, all students strongly agreed that the course had allowed more direct interaction with lecturers and researchers than had occurred in their previous year of university study. They also reported that they had learned more than in lectures or laboratory practicals, that they had been able to apply their learning immediately and that they had enjoyed meeting new people and working with others. The challenges were, of course, evident, but for the most part students acknowledged these positively:

> Collaborative projects are challenging in themselves, but it was definitely the immense pressure to perform in a small amount of time that was the most difficult. I think this was helped by the great support and assistance of the resource people, and the general enthusiasm/dedication of all of the students on the course.

The responses to open-ended questions—which repeatedly showed that the learning goals of the course had become the learning outcomes of its students—often provided unexpected joy in the hearts of the design team as educators and scientists:

> The field course definitely has changed my perception on research … [It's] not just about discovering new things, although it is an exciting part. It is also about confirming already established ideas, or disproving them and creating new theories.

***

> From the lectures to the practical fieldwork, this course has been an experience I will never forget. I do not regret anything about it and look forward to applying all the skills I have picked up.

In summary, the 2015 inauguration of the *Field Studies in Functional Ecology* course at The Australian National University was a great success. It also provided the members of the design team—self-identified as lifelong learners—many opportunities to tweak and improve specific elements.

This has been done in the subsequent successful iterations of the course—again at Charlotte Pass in 2016 and, taking the concept to tropical north Queensland, in the Daintree rainforest in 2017. Interestingly, several of the 2015 students chose to have a second bite of the cherry, which required another level of thought around appropriate learning outcomes for those doing the course for the second time. Future volumes in this series will document the research findings of those later scientific adventures.

It is fitting, however, for the final words of this volume to come from a 2015 student, whose reflective feedback epitomises the research-focused outcome that the design team initially envisaged, and that made the many, many exhausting days and weeks spent on planning, executing and evaluating this course so worthwhile:

> I believe that I now am more capable at looking at the world scientifically and seeing questions where I used to just see curiosities.

References

Beavis, S, Beckmann, EA (2012) Designing, implementing and evaluating a consultancy approach to teaching environmental management to undergraduates. *International Research in Geographical and Environmental Education* 21(1): 71–92. doi.org/10.1080/10382046.2012.639151

Durrant K, Hartman TPV (2014) The integrative learning value of field courses. *Journal of Biological Education* 49(4): 385–400. doi.org/10.1080/00219266.2014.967276

Haslam, SA (2004) *Psychology in Organisations: The Social Identity Approach*, 2nd edn. Sage, London.

Healey, M (2005) Linking research and teaching: Exploring disciplinary spaces and the role of inquiry-based learning. In Barnett, R (ed.), *Reshaping the University: New relationships between research, scholarship and teaching*, pp. 67–78. Society for Research into Higher Education and Open University Press, Maidenhead.

Honebein, P (1996) Seven goals for the design of constructivist learning environments. In Wilson, BW (ed.), *Constructivist Learning Environments: Case studies in instructional design*, pp. 17–24. Educational Technology Publications, New Jersey.

Howitt, S, Wilson, A, Higgins, D (2014) *Teaching research: Evaluation and assessment strategies for undergraduate research experiences (TREASURE)*, Final Report 2014. Australian Government Office for Learning and Teaching, Sydney.

Moon, J (1999) *Reflection in Learning and Professional Development: Theory and practice*. Routledge-Falmer, Abingdon.

www.ingramcontent.com/pod-product-compliance
Lightning Source LLC
Chambersburg PA
CBHW042320210326
41599CB00048B/7178